The Culture of Military Innovation

The Culture of Military Innovation

THE IMPACT OF CULTURAL FACTORS ON

THE REVOLUTION IN MILITARY AFFAIRS

IN RUSSIA, THE US, AND ISRAEL

Dima Adamsky

Stanford Security Studies
An Imprint of Stanford University Press
Stanford, California

Stanford University Press
Stanford, California

Special discounts for bulk quantities of Stanford Security Studies are available to corpora-
tions, professional associations, and other organizations. For details and discount information, contact the special sales department of Stanford University Press. Tel: (650) 736–1782,
Fax: (650) 736–1784

Printed in the United States of America on acid-free, archival-quality paper

Library of Congress Cataloging-in-Publication Data

Adamsky, Dima.
The culture of military innovation : the impact of cultural factors on the revolution in
military affairs in Russia, the US, and Israel / Dima Adamsky.
    p. cm.
  Includes bibliographical references and index.
  ISBN 978-0-8047-6951-8 (cloth : alk. paper)—
  ISBN 978-0-8047-6952-5 (pbk. : alk. paper)

  1. Soviet Union—Armed Forces—Technological innovations. 2. Military art and
science—Technological innovations—Soviet Union. 3. Strategic culture—Soviet Union.
4. United States—Armed Forces—Technological innovations. 5. Military art and
science—Technological innovations—United States. 6. Strategic culture—United States.
7. Israel—Armed Forces—Technological innovations. 8. Military art and science—
Technological innovations—Israel. 9. Strategic culture—Israel. I. Title.

UA770.A57 2010
355'.07—dc22   2009021452

Typeset by Westchester Book Group in 10/14 Minion

*To my grandfathers, Jewish officers
in the Soviet Army during
the Great Patriotic War, whose
examples always inspired me.*

CONTENTS

ACKNOWLEDGMENTS

THE RESPONSIBILITY for the views and mistakes of this book is mine alone. The better parts of this work would not have been possible without the generous assistance of many people, and I would like to briefly express my debt of gratitude to them here.

It was Uri Bar-Joseph of Haifa University who grounded my professional fascination with defense transformations in academic research, during our discussions about international politics and modern strategy in one of the finest coffeehouses in Tel Aviv. From my disorganized and associative thoughts, he developed a coherent observation about the puzzling intellectual history of the RMA in the USSR, the US, and Israel. Explaining variation in innovation across military organizations became the essence of my dissertation and the topic of this book. I owe him my deepest gratitude for the idea of this work, for his devoted supervision, and for several other important decisions that I have made. Each student and practitioner of international security dreams of having this kind of superb academic mentor. Ben D. Mor suggested the conceptual framework suitable to deal with this empirical puzzle and introduced me to the world of IR theories. My research would not have been possible without the academic assistance of Gabriel Ben-Dor and Abraham Ben-Zvi, all of them from the School of Political Sciences at the University of Haifa.

In the middle of working on this project, I was exceptionally fortunate to relocate to the John M. Olin Institute for Strategic Studies at Harvard University. Intellectual life at Harvard opened unparalleled opportunities for enhancing theoretical and empirical knowledge. It was an academic honor and intellectual privilege to be supervised by and to learn from Stephen Peter Rosen, director of the institute, who by virtue of his example educated us to the highest standards of social science research and to think outside of the box with scientific elegance. It would be an understatement to say that I am grateful to him for many opportunities. I was lucky to benefit from the open door policy of Mark Kramer of the Davis Center at Harvard, whose knowledge of Russian and Soviet military affairs is encyclopedic and whose reflections are always invaluable. The advice of Alastair Iain Johnston and conversations with Jacqueline Newmyer have significantly contributed to my understanding of how to apply cultural and ideational analysis to security issues. For their comments, assistance, and advice, I am grateful to the fellows of the Olin Institute, particularly Lindsay Cohn, Jeffrey Mankoff, Assaf Moghadam, and Keren Yarhi-Milo. For one year, I was lucky to inhabit an office with Col. Michael Hays, who shared with me the perspectives of a bookish combat pilot on many relevant issues and provided "close air support" each time I called for it.

I wish to acknowledge the most valuable input and remarks that I have received from Andrew J. Bacevich, Brig. Gen. Itai Brun, Michael Horowitz, David Johnson, Mary C. FitzGerald, Andrew F. Krepinevich, Andrew May, Brig. Gen. H. R. McMaster, Lt. Gen. William Odom, Michael G. Vickers, Barry Watts, and Vladislav Zubok. Azar Gat of Tel Aviv University and Theo Farrell of Kings College London graciously shared with me their outstanding expertise and offered invaluable constructive suggestions. I have tried to accommodate those in polishing my argument and improving the final product, but probably much is still left to be done. My thanks are also owed to acting senior officials and military officers in the three countries who provided nonattributed remarks and advice.

The curriculum of the SWAMOS workshop on Military Operations and Strategy at Cornell University, and the mentorship of the leading scholars in this field, provided me with important methodological and analytical tools. I have learned a great deal about change in the nature of modern warfare from Eliot A. Cohen, who kept on mentoring me whenever I needed his academic guidance, professional advice, and help during those years. I have learned as much about continuity in modern warfare from Stephen Biddle, who kindly provided me with his uninterrupted academic support. In addition to the superb lessons in utility of force during SWAMOS, I am much indebted to Richard K. Betts, director of the Saltzman Institute of War and Peace Studies, for hosting me as a visiting researcher at Columbia University during the final stage of this work. His academic hospitality, support, and advice enabled me to concentrate on reworking the manuscript for publication.

I want to thank Minister Dan Meridor for the fascinating lessons of how to think about international security with breadth, depth, and integrity. Professional experience with this true defense intellectual significantly enriched my academic perspective on how defense bureaucracies and military machines function from within.

I am indebted to Yaacov Ro'i of Tel Aviv University for his confidence and encouragement before and during my stay at Harvard. A special word of thanks is due to Deena Leventer, who was always there for me as a friend and as a professional. Her work is unparalleled in many ways, but especially because it enables me to realize what my views really are and are not. I benefited enormously from the advice and assistance of Aviva Cwang, the chief librarian of the IDF General Staff Library. I owe my appreciation to Malcolm Byrne, William Burr, and Svetlana Savranskaya at the National Security Archive for their help and guidance. Portions of this book are based on articles previously published in the *Journal of Strategic Studies* and in *Defense and Security Studies*. I gratefully acknowledge the journals' permission to rework this material and particularly want to thank the editors: Therese Klingstedt, Thomas G. Mahnken, and Joe A. Maiolo. I want to thank Jen Sundick for her devoted and excellent editing and to extend my appreciation to Geoffrey RH Burn, director and acquisitions editor of the Security Studies Series, and Jessica Walsh, editorial assistant at Stanford University Press, for their most thoughtful and personal approach. My deep gratitude to Nilly Sikorsky for her generous support of my research.

While working on the project I spent a lot of time in each of the case countries. Friends and colleagues assisted me, each one in a special way, and I owe a debt of gratitude to each and every one of them. My dearest friend Maxim Safyan (aka Gap) never understood exactly what I was writing about and why, but he always fulfilled my endless requests, collected half of the sources for this book, supported me during the most difficult moments, provided great suggestions, and inspired me with his beautiful mind, one of the sharpest I have ever known. Many ideas crystallized during intellectual exchanges with Noam A., Nehemia Burgin, Col. Ron E., Ehud Eiran, Michael Gurevich, Col. Yoram H., Miki Haimovich, Doron Horesh, Yaacov Falkov, Shai Granovsky, Dudi Malik, Gai M., Nuno Monteiro, Margaret Sloane, Aviad Sella, Deganit

Paikowsky, Ely Ratner, and Daniel R. This book, like many other projects, would not have been possible without the technological genius of Vadim Punsky, who has literally resurrected the project several times. I also want to thank Joe Gehr for assisting me in picking the title.

Last, but certainly not least, I want to thank Ania, who realized how important this project was for me and tolerated my often intolerable and awkward behavior. Her brilliant remarks made complicated issues simpler and highlighted the complexity of those issues that initially looked straightforward. Her participation was indispensable.

The Culture of Military Innovation

INTRODUCTION

THIS BOOK studies the impact of cultural factors on the course of military innovations. It shows to what extent different strategic cultures—specifically the Soviet, the American, and the Israeli—account for the varying approaches to transformation in the nature of war. Revolution in Military Affairs (RMA) and Military-Technical Revolution (MTR) are the terms used for a radical military innovation, in which new organizational structures together with novel force deployment methods, usually but not always driven by new technologies, change the conduct of warfare.[1] The term "revolution" does not mean that the change will be rapid but implies that it will be profound, making new methods of warfare more powerful than the old.[2] Consequently, RMAs matter much because they render the existing forms of combat obsolete.[3]

Most military revolutions have arisen from technological advances. However, RMAs are driven by more than breakthroughs in technology, which in themselves do not guarantee successful innovation. "Technology only sets the parameters of the possible and creates the potential for military revolution. What indeed produces an actual innovation is the extent to which militaries recognize and exploit the opportunities inherent in new tools of war, through organizational structures and deployment of force. It was how people responded to technology that produced seismic shifts in warfare," argues Max Boot, who inquired into the nature of military revolutions since 1500.[4] History is full of examples to support this claim.[5] While the technological component is often an important initial condition, a true revolution depends on confluence of weaponry, concept of operations, organization, and the vision of future war.[6] According to Andrew W. Marshall, the director of the Office of Net Assessment in the Pentagon and one of the driving forces behind modern American strategic thought, "the main challenge in the RMA is an intellectual and not a technological one."[7]

The earlier defense experts recognize and understand the discontinuity in the nature of war, the better. Foreseeing an RMA is not a talisman for military victory. That being said, it is highly probable that anyone who anticipates the RMA and transforms their forces accordingly will significantly enhance military effectiveness.[8] A delay, consequently, will have the reverse effect. The price of mistakes can vary from battlefield ineffectiveness with minor tactical implications to strategic catastrophe with devastating consequences for national security.[9] There were cases in history when, early on, the significance of unfolding RMAs was recognized,[10] but for the most part RMAs have been acknowledged only after the fact.[11] Recognition of an RMA and the concomitant development of a vision of future war is an act of unquestionable importance. The need to anticipate the trends of the next conflict, and not to prepare for the previous one, was adopted by defense analysts as their mantra. The ability to diagnose and to understand the discontinuity in the nature of war— the rapid change in ways and means of fighting—is probably the most critical aspect of defense management. Imagining the future enables defense managers to embark in real time on crafting of what Stephen Peter Rosen calls the "new theory of victory."[12]

THE PUZZLING HISTORY OF THE CURRENT RMA

Since the early 1990s, the US and other world militaries have argued that the most dramatic revolution in warfare since the introduction of nuclear weapons is under way. In mechanical terms, the Information-Technology Revolution in Military Affairs (IT-RMA) integrated long-range, precision-guided munitions (PGMs), C4I (command, control, communications, computers, and information), and RSTA (reconnaissance, surveillance, targeting acquisition) in a form that completely changed the combat environment and altered the way militaries think about the aims and methods of conventional warfare. In terms of basic capabilities, the IT-RMA entails the ability to strike with great accuracy, irrespective of range; the ability to penetrate defensive barriers using stealth technology and unmanned warfare; and the ability to move information rapidly across a joint battle network and exploit the effects of increased joint force integration.[13]

In terms of organizational structures and concepts of operations, classical patterns of advancement along fronts with discernible lines and rear areas have disappeared; the number of platforms has become far less important than networks and communications; military planning aims at defined effects rather than attrition of enemy forces or occupation of territory; instead of massive forces, precise fire is maneuvered; the sensor-to-shooter loops have been shortened considerably; the role of stand-off and airpower capabilities have increased at the expense of heavy ground formations; and a far smaller, lighter, and more mobile force can operate on a greater range and with higher precision and greater lethality than at any time in human history.[14]

The 1990 Gulf War offered for the first time a glimpse at the revolutionary potential embodied in these various combat capabilities. Nearly a decade later, in 1999, Allied operations in Kosovo reinforced the value of so-called information warfare for future military campaigns. Operations in Afghanistan and Iraq in 2001 and 2003 have demonstrated how conventional wars can be fought using this "high-tech, low numbers" operational approach.[15] The principles underpinning the RMA diffused around the globe and created a consensus regarding the characteristics of a modern conventional military: a small, highly skilled joint force, versatile for conventional warfare and counterinsurgency, flexible, rapidly deployable, and using advanced information technology.[16]

The roots of this RMA can be traced to the mid-1970s, when stand-off, precision-guided munitions were introduced in military theory and practice in the USSR, the US, and Israel. When the US shifted its attention away from Indochina and back to the European theater, it faced a daunting strategic situation: increasing realization of mutually assured destruction eroded the option of nuclear war; yet, in the balance of conventional forces, the Soviets were predominant. To destroy the menace of Soviet second echelons deep in the rear, the US capitalized on stand-off precision-guided munitions (PGMs) and "over-horizon" sensors, based on state-of-the-art scientific-technological developments in the field of microelectronics.

History knows many examples where the nation other than the inventor of the technology was first to understand its potential and to exploit it on the battlefield.[17] Although it was the US that was laying the technological groundwork for the RMA, Soviet rather than American military theorists were the first to intellectualize about its long-term consequences. In contrast to the West, which focused on the weapons' narrow implications, the Soviets were pioneers in championing the argument that the new range of technological innovations constituted a fundamental discontinuity, which they dubbed the "Military-Technical Revolution." Beginning in the late 1970s, the Russians produced a significant number of seminal works on the MTR. They actually predated the West by almost a decade in realizing the revolutionary essence embodied in US and NATO military-technological shifts. The Soviets used the West's scientific and technological superiority as a conceptual starting point for their own doctrinal innovations. For political, economic, and cultural reasons, the gap between MTR conceptualizations and the actual capabilities of the Soviet military forces was never bridged. Later, Soviet theories provided a looking glass for US strategists. From the early 1990s, following almost a decade of Western conceptual disregard, a fundamental Soviet MTR vision was analyzed and adapted by the US. American defense theorists coined the term "Revolution in Military Affairs" after adopting the Soviet variations on that theme. It was by studying the reflections of American military power in Soviet eyes that American strategists fully comprehended in the early 1990s the significance of what they had produced.[18]

While the Americans were the first to produce RMA-type munitions, and the Soviets were the first to theorize on their revolutionary implications, the Israeli Defense Forces (IDF) were the first to employ these revolutionary weapons in combat. After the 1973 Yom Kippur War, the IDF intensified their investment in PGM weaponry, sophisticated "over-horizon" intelligence capabilities, and command and control systems. Although Israel was the first to make extensive use of IT-RMA technology on the battlefield in Lebanon's Bekaa Valley in 1982, Israeli defense specialists viewed precision-guided munitions simply as another force multiplier in the IDF's conventional arsenal. The conceptual and doctrinal emulation of the US since the mid-1990s touched off broader changes in IDF thinking, introducing the aspiration to be a "small and smart military," à la RMA.

The intellectual history of the RMA is puzzling because of the diverse paths that led each nation to the same military innovation. At the outset, the US developed technology and weaponry for about a decade without realizing their revolutionary implications. No attempt to reconceptualize the existing paradigm about the nature of warfare in futuristic terms was made by the US in those years. Soviet MTR concepts represented a form of theoretical conceptualization that chronologically preceded technological procurement and combat experience. The Soviets were the first to speak of this development as a discontinuity, coined the term "Revolution in Military Affairs," and produced seminal theoretical works without possessing either weapons or technologies. The Israelis utilized the weaponry on the battlefield, but only ten years later, beginning in the mid-1990s, did the IDF start to reform its concept of operations, force structure, and broader strategic doctrine to fully exploit the opportunities provided by information technology. The first country to extensively use RMA technologies in combat, Israel was the last to develop a conceptual framework that acknowledged their revolutionary implications. Similar to the American case, the cultivation of the technological seeds of the Israeli RMA preceded the maturation of the conceptual ones.

This process is counterintuitive; one would expect that countries accustomed to similar technologies would undergo analogous and simultaneous paradigmatic changes in the perception of warfare. However, in the cases in question, the evidence indicates that this is not necessarily so: enormous variation is evident in the intellectual paths taken toward conceptual innovation. Moreover, the gaps of time between familiarity with the technology and systematic understanding of its potential implications for organizational structures and concepts of operations vary considerably across the cases. How can disparate intellectual behavior of military theoreticians and civilian strategists in the USSR, the US, and Israel with regard to the RMA be explained? Why did the technologically inferior USSR prevail qualitatively over the US and Israel in the conceptualization of the emerging military-technological realities? Why did it take the US defense community close to a decade to acknowledge the accuracy of Soviet assumptions and to translate MTR theoretical postulates into a radical military reform? Why did Israel not embark on the transformation of its military much earlier? The conspicuous variation in intellectual approaches generates two central questions of this book: What explains the different ways in which military innovations, based on similar technologies, develop in different states? How does a "new theory of victory" originate in a certain ideational milieu, and what contributes most to shaping its nature?

CONCEPTUAL FRAMEWORK

Scholars of military innovations maintain that Revolutions in Military Affairs depend as much upon gaining access to the requisite technologies as on restructuring concepts and organizations. In the last two processes social and cultural factors are critical. This book, which relates to military innovation studies, will elaborate on various factors in the field that have, to date, been identified as shaping innovation. A body of literature about the imprint of cultural attributes on strategic behavior provides the most relevant answers to the questions posed in this book.

Chronologically, the works about the cultural impact on national security policy, which were introduced under the umbrella rubric of "strategic culture," came in three waves. The first generation of scholarship emerged in the late 1970s and focused on the link between national political and military cultures, and the strategic choices that countries made. The literature argued that a deeply rooted set of beliefs and a nation's formative historical experiences create its distinct mode of strategic thinking and particular attitude toward security affairs. The notion that different security communities might think in different ways about the same strategic matters began to gain acceptance.[19] Empirically, the literature concentrated mostly on the distinctive national styles in the superpowers' grand-strategy making and on the cultural roots of the nuclear doctrines of the USA and the USSR.[20] The second wave of literature came in the early 1990s and presented strategic culture as an independent determinant of security policy patterns. A variety of case studies sought to prove that "strategic culture" constituted the milieu within which strategy was debated, and if not ultimately driven by the parameters of strategic culture, national security policy had deep cultural underpinnings.[21] The literature of this period sought research methodology to make the analytical models of the discipline less opaque, vague, and simplistic.[22] The third generation of scholarship on the cultural foundations of strategic behavior picked up in the mid-1990s. The constructivist research program, which emphasized the ideational construction of international politics, "devoted particular attention to identity formation, the organizational process, history, tradition, and culture," and provided a far more nuanced picture of security affairs. It took researchers of military innovation deeper within states, organizations, and the process of producing new technology, to account for the role of culture and norms.[23] Constructivists framed the impact of domestic factors on security as a coherent theoretical paradigm to counterbalance the traditional neorealist approach to international security.[24] Constructivism views culture as a synthesis of ideational meaning that governs perceptions and actions, defines the situation, articulates motives, and formulates a strategy for success.[25]

In the last decade, a growing interest in ideational explanations resulted in numerous studies of a variety of empirical and theoretical topics that all appeal to socially, culturally, and ideationally independent variables to explain military affairs. A number of proponents of the cultural approach to IR sought to inquire into the ideational foundations of states' defense policy and concentrated on the interrelation between norms, culture, and strategic behavior.[26]

Other scholars focus on domestic social structures to explain particular national styles in generating military power.[27] The renewed interest in organizational analysis led to an appreciation of the interstate level and to focusing on "figuring out the fighting organizations."[28] Scholars paid considerable attention to the linkage between the nature of the organization and the military innovation it produced.[29] At the intrastate level, scholars examined how the cultural identities of specific nations shape military doctrines,[30] and paid attention to the intellectual dynamics and adaptive learning between institutions from different states that influence the preferred paths of strategic behavior.[31]

Inspired by constructivism, the literature on the impact of culture on military innovations spanned beyond the traditional political science and has been informed by cross-disciplinary linkages to anthropology, historical research, sociology, and psychology. Concurrently, a belief that military reality is socially constructed evolved among historians.[32] Social military historians argued that collective memoirs and imagination shape the way nations prepare for, conduct, and experience warfare. They provided useful insights into how collective memory and imagination of war are framed[33] and how this memory and these images shape social identity[34] and the state's approach to warfare and security.[35] This research draws upon the assumption that "cultural construction" is the best means of understanding the differences between militaries.[36] Literature on the social approach to technology studies is another useful source of insight. It envisions technological innovation as ideas shaped by discourse between "interest groups" that view the potential of a given technology differently, in a manner that corresponds to their beliefs, preferences, and vision of what that technology can do for them.[37] In this way, military technologies are not deterministic, but socially constructed.[38]

In his superb review of military innovation studies, Adam Grissom divides the literature of the field into distinct categories according to the focus of analysis. The six avenues of research comprise the civil-military, intraservice, interservice, cultural, top-down, and bottom-up models of military innovation.[39] This book refers to several categories of this typology across cases. However, according to Grissom's definition this study would be qualified as part of the cultural school of research, in the context of the major works in the field of military innovation. This work further elaborates on the recent scholarly definition of "strategic culture," which views it as a set of shared formal and informal beliefs, assumptions, and modes of behavior, derived from common experiences and accepted narratives (both oral and written), that shape collective identity and relationships to other groups, and which influence and sometimes determine appropriate ends and means for achieving security objectives.[40]

LACUNAE OF KNOWLEDGE

Notwithstanding the established contributions of the literature, empirical and theoretical lacunae emerge from the existing body of research.

Literature related to IT-RMA in the USSR, the US, and Israel may be broadly divided into two complementary and often overlapping categories: practice-oriented research on the nature of the RMA, and research about the history of the RMA development. The studies belonging to the first category include empirical data and theoretical observations that then guide the community of experts involved in defense transformation policy making. Studies belonging to the second category consist of subjective interpretations and observations of the RMA as a stage in the intellectual development of modern military thought. In each of the three cases under study, the practice-oriented body of literature is considerably larger than the intellectual history of the RMA. Most of the studies that concentrate on the history of the Soviet MTR are a priori deficient, inasmuch as they were produced during the 1980s or early 1990s, when their authors had no access to critical collections of declassified Soviet materials. The American case has been dealt with extensively, but there is still a conspicuous lacuna regarding the role of intellectual emulation of the Soviet MTR and regarding the diffusion of military ideas from West to East and back again. The history of Israeli military thought in the last decades in general, and the history of the IDF RMA in particular, remain, for the most part, unwritten. The few existing sources are full of caveats and lack in-depth perspective.

Though various studies portray how military innovations, and particularly Revolutions in Military Affairs, develop through organizational, doctrinal, or technological processes, the researchers only recently started to focus on the symbiotic relationship between technology, strategic culture, and development of the new theory of victory.[41] Strategic culture literature suffers from an acute deficit of comparative studies. As a rule, scholars focus on the impact of strategic culture on the behavior of one or two actors, while inquiring into a particular issue of international security—usually a nuclear policy. The literature has paid attention mostly and primarily to the American traits of strategic behavior. Relative to the American case, Soviet-Russian strategic culture has been neglected. With regard to Israeli strategic culture, the literature is completely silent.

A few researchers of strategic culture have utilized social sciences beyond IR.[42] Anthropologists, sociologists, and psychologists agree that culture conditions behavior and thinking style. The latter insight is directly relevant to, but underdiscussed in, security studies. Cognitive style, which is an essential part of a state's strategic culture, remains outside the scope of academic investigation. There appear to be no studies to date that examine military innovations through the lens of cultural psychology and observe the impact of cognitive styles on developing a new theory of victory. While security studies scholars have overlooked the framework of analysis offered by the cognitive style theories, this approach is widely applied in cross-cultural psychology.

To sum up, despite the important pioneering role the USSR, the US, and Israel have played in shaping current military innovations worldwide, there has been no systematic, cross-cultural, comparative analysis of the intellectual history of the RMA in the three countries to date. This book seeks to fill the above-mentioned theoretical and empirical voids. It suggests a cultural explanation for significant incongruence in approaches to this military innovation in the US, the USSR, and Israel.

RESEARCH OBJECTIVE

The book sheds light on the sources of the strategic cultures and portrays the determinants of a conceptual approach to military innovations. It argues that, given the initial and necessary conditions provided by technology, the variation in the military innovation in the USSR, the US, and Israel was ultimately shaped by the impact of each state's strategic culture. The book addresses the cultural and cognitive factors that affected the dependent variable, presents the causal mechanism that sets the relationship between culture and generation of an RMA into motion, and inquires into a host of questions regarding the impact of norms on developing new theories of victory.

The research does not pretend to seek a positivist wisdom that will enable unconditional predictions about strategic behavior of the three countries under study; neither does it attempt to develop a universal explanation that can clarify the impact of cultural factors in all cases of military innovations. However, it seeks to show the persuasive plausibility of the causal mechanism and to improve scholarly understanding of the relationship between culture and military innovations, capitalizing on the intellectual history of the recent RMA. Using comparative and contrasting cases, this study produces more credible evidence for the influence of cultural factors than can an example in which only one society is being considered. It focuses on the fundamental strategic narratives and traditional military beliefs that shape the intellectual climate of a security-making community and its approach to military innovation. The book shows how thinking styles came about in each case study, and how they set each state's analytical approach to the RMA into motion.

The question of whether strategic culture is the cause or the context of action goes to the heart of the once vigorous theoretical debate. Indeed, the mere act of defining strategic culture as an independent variable remains controversial, and research may border on tautology, whereby sociocultural structures are seen as both reflecting and shaping behavior.[43] This book focuses on sociocultural factors to account for the variations of the dependent variable. Yet it echoes the approach of those who see strategic culture as an ideational milieu that frames behavioral choices, and defines it as an intervening variable. It remains true to the claim that culture influences action, "by shaping a repertoire of skills, styles and practices from which people can construct strategies of action."[44]

Ontologically, emerging technologies and strategic culture are not independent of each other but interact in the context of organizational and strategic environments. The relationship between technology and military innovation is not deterministic, but rather socially constructed; national military tradition and professional cultures interact with technology, affecting the course and outcome of military change. The kind of weaponry that is developed and the kind of military that it foresees are cultural products in the deepest sense.[45] A true revolution in the way military institutions organize, equip for, train for, and conduct war depends on a confluence of political, social, and technological factors.[46] Each factor is necessary but insufficient for paradigmatic change. An interaction among a number of variables (structural, technological, and cultural) sets off a complex chain reaction that generates military innovation, in a particular chronological order. Structural factors and emerging technologies represent the independent starting point of military transformation. The cultural factor is not, in itself, sufficient for the emergence of the RMA but constitutes the pivotal intervening variable that conditions innovation's path of development. Consistent with this argument, this book chooses the cultural factor as its main focus and takes up the discussion from the point at which the technology has been introduced. Although the breakthrough technology usually constitutes a basic precondition for paradigmatic change, the cultural factor is critical for assessing the causal effects of the hypothesis. "New theory of victory" is a quintessentially cultural-ideational endeavor. Cultural factors explain why, once a new technology is available, certain states translate it into an RMA while others do not, or do it in a different style. Thus, with "knowledge of a technology" held constant across the cases, ideational differences account for variance in the course of the development of a new theory of victory. It follows that states may utilize the same technologies in different ways, in accordance with their sociocultural patterns, and eventually produce different military innovations.[47]

Structural and material factors come to mind as the main alternative explanation of the Soviet, American, and Israeli approaches to the RMA. The proponents of this explanation would argue that the need to survive forced the above three states to prepare for war as efficiently as possible, under the objective material limitations. As Kenneth Waltz sees it, states generate maximum military power using available national potential and attempt to "mimic the military innovations of countries with greater capabilities and ingenuity."[48] Barry Posen echoes this claim, positing that in any competitive system, successful practices will be imitated.[49] Thus, according to this explanation, each state responded to changing strategic circumstances, to the changes in threats, and to the balance of power. The USSR, the US, and Israel operated along rational lines, made the best use of available resources, emulated each other's practices, and sought to organize their military forces in the most functional manner. The scope of the military innovation in each country was conditioned by the resources available to each of the innovators.

However, empirical evidence runs counter to this explanation. Behavior of each state seems dysfunctional, if not irrational. Why did the Soviets develop doctrines incompatible with the country's capacity to implement them? Why did the US and Israel produce and utilize the weaponry but not embark on the innovation of their militaries much earlier? Neorealism is weak in explaining this counterintuitive strategic behavior. An ideational approach, which argues that the perception of security interests is a function of a country's cultural environment,[50] offers a better-suited analytical framework for this task. Cultural analysis is not a substitute for but an effective supplement to neorealism for explaining states' strategic behavior.[51] It does not reject neorealistic claims about rationality or lack thereof of the actors. Instead it argues that rationality is neither objective nor universal, and that rational behavior is culturally dependent. The actors formulate their preferences not in accordance with a universal logic of efficacy, but according to their own norms, values, and self-image.[52]

METHODOLOGY

To show how cultural variables form the ideational framework for military innovation, this study draws upon theoretical insights and models of analysis from political science, history, sociology, cultural anthropology, and cognitive psychology. All these approaches have a large measure of validity and are not, for the most part, mutually exclusive, but are reciprocally complementing.[53]

According to Jeffrey Lantis, strategic culture research lacks a defined methodological design.[54] Studies in the field frequently have difficulties in illuminating variables that constitute the universal parameters of "strategic culture." In part this reflects differences among the cases: certain organizational or ideational parameters that play a significant role in one country may play a secondary role in another. The common solution, adopted also by this book, is to choose according to the "lasting nature" of the impact. Features of culture, norms, and ideas that transcend generations and impact continuously upon a state's strategic behavior are chosen as the parameters of the strategic culture in a given case. To keep the research model elastic but not to deteriorate to parsimonious definition, scholars choose strategic culture parameters from three interdependent pools: national-popular culture, characteristics of policy-making mechanisms in security affairs, and "bureaucratic reflexes" and organizational cultures of defense institutions.[55] In compliance with this methodological approach, and in keeping with the topic of the book, this research includes the following factors under the rubric of parameters of strategic culture: social structure; cognitive style; cultural time-orientation; styles in communication; organizational approach to innovations; preferences in ways of waging wars; structure of the military system and the role of the General Staff, the Joint Chiefs of Staff, and the General Headquarters in it; approach to the development of military knowledge; the role played by technology in military affairs; and the approach to weapons procurement.

Determining strategic traditions entails a thorough analysis of primary sources.[56] This book is based on data obtained from archival documents, oral testimony and memoirs of relevant insiders, curricula of military academies, field and training manuals, and leading professional military periodicals. The

Soviet case is constructed using declassified dissertations and scientific studies written by senior Soviet military officers, as well as declassified military periodicals and theoretical writings that formed the basis of the MTR paradigm. In the case of the US, the study utilizes declassified documents produced by the defense and intelligence communities that reflect the adaptive learning and construction of the RMA concept by the American defense establishment. The Israeli history of the RMA is constructed using the curricula of military academies, leading professional journals, and interviews. In the cases of the USSR and Israel, the study is based on primary sources that had previously been completely inaccessible. Where no available primary materials exist, the study relies upon secondary sources. These are employed not only as a substitute for declassified records, but as an important means of illustrating the cultural context that shaped the decision-making process.

Following the introduction, the book comprises five chapters: the relationship between cognitive styles and understanding revolutions in military affairs is discussed in the first chapter; the following chapters address the Soviet, the American, and the Israeli cases. Each case study starts with the empirical narrative about the course of the RMA. It is followed by the portrait of each respective country's strategic culture and the cognitive style of its military experts. The conclusion of each chapter analyzes the impact of cultural factors on the course of the innovation and illuminates the causal link between the two. The conclusion of the book synthesizes insights from the three cases and discusses the intellectual history of the RMA from a cross-cultural perspective. As an integrated whole, it summarizes the findings on the RMA's development in each of the three cases; analyzes the role of cultural variables in shaping the variations and the congruencies of the military innovation; highlights directions of future research; and elaborates on policy implications.

Several important questions related to the IT-RMA have intentionally been left outside the scope of the current study. The discussion deliberately refrains from addressing the question of whether the process described actually represented a revolutionary discontinuity in warfare, and it does not examine the efficacy of the IT-RMA-type warfare against asymmetrical threats.[57] It does not explore the debate brought by the "military reform" movement,[58] does not deal with the Soviet reaction to the US Strategic Defense Initiative,[59] and does not discuss the way in which nuclear doctrines are influenced by cultural variables.

Though framed in the context of Soviet, American, and Israeli historical experience, the insights of this book will be of wider importance for practitioners, scholars, teachers, and students of security studies. The book seeks to make a threefold contribution. Historically, the research bridges the gap between the significance of the IT-RMA for the development of modern military thought and the poorly grounded academic debate on its intellectual history. Theoretically, it generates a set of hypotheses to guide a cultural approach to security studies and provides a fresh, comparative look at theories of military innovation. In practical terms, the cultural analysis presented in this book provides defense managers with tools to assess the weaknesses, strengths, and behavioral inclinations of friends and foes and enables them to understand the foreign lenses through which one's policy is grasped. A thorough understanding of the strategic cultures of others contributes to formulating policy in the fields of deterrence, bargaining, and strategic signaling and increases accuracy during intelligence analysis of foreign operational behavior and strategic intentions.

1 COGNITIVE STYLES AND UNDERSTANDING REVOLUTIONS IN MILITARY AFFAIRS

INTRODUCTION

"Good strategy presumes good anthropology and sociology," argued Bernard Brodie in his *War and Politics*.[1] In other words, good anthropology and sociology are tools of great significance for understanding how strategy is crafted and how military experts tailor a "new theory of victory." One can make an argument that variance in strategic behavior stems from differences in strategic cultures. In turn, differences in strategic cultures are determined by, among other factors, the variance in fundamental cultural characteristics.

A national cognitive style is one element in the cultural mosaic that shapes a state's strategic behavior and constitutes the ideational foundation of its military innovation. Empirical evidence gives grounds to assume that experts in the same profession from different cultures think differently about military innovation and produce various types of doctrinal outcomes from the same technological discontinuity. What reasoning traits recur in the intellectual activity of military experts in the frame of a given culture? What factors shape, determine, and guide experts' imaginations and cognition when they conceptualize military innovation? This chapter addresses these questions and portrays how the sociocultural background of experts shapes their style of reasoning. It introduces several fundamental cultural distinctions among societies, with particular focus on the *cognitive style*. This focus aims to present the plausible link between the cultural factors of a given nation and its aptitude to understand revolutions in military affairs. I argue that given certain cultural and cognitive characteristics, a particular professional community may have a greater propensity than others to grasp paradigmatic change in the nature of war.

The chapter elaborates on this argument in three steps. The first section describes a variance among different cultures, by distinguishing them according to *social structure* (collectivistic vs. individual), *communication style* (high context vs. low context), and *time orientation* (polychronic vs. monochronic); then, the second section discusses how these cultural differences underlie inclinations in *cognitive styles* (either towards holistic-dialectical cognitive style or towards logical-analytical reasoning); finally, the third section explains why cultures that are inclined towards holistic dialectical reasoning may have greater aptitude for understanding the RMAs. The description of each country's strategic culture, which will follow in the subsequent chapters, will make reference to these cognitive characteristics.

SOCIOCULTURAL DIFFERENCES AND COGNITIVE STYLES

The quintessence of the general cultural traits of a given country impacts national strategic behavior. Theories of cultural psychology (sometimes referred to as anthropological or ethnic psychology) arose from the question of whether the laws that govern the behavior of an individual could be extrapolated to the macro-level to account for the functioning of a specific society as a whole. The predominant tendency is to combine psychology, which deals with micro-level phenomena whose scope in time and space is limited, with anthropology, which tends toward macro-level concentration on history and culture.[2] Cultural psychology suggests distinguishing between societies by situating them along several cultural and cognitive continuums.

Cultural Continuums

Since different cultures transmit information differently, the literature suggests distinguishing between cultures by situating them along a dimension spanning high-context to low-context communication styles. In this framework of analysis, introduced by Edward Hall, the context referred to consists of background, the relationship between interlocutors or environment, and context connected to the discussed focal event or topic. An individual from a high-context culture expresses himself or herself (orally and in writing) in indirect, reserved, cyclical, and vague language, relying on the listener's/reader's ability to grasp the meaning from the context. This type of culture conveys information through messages with deep meaning and without referring to the problem directly. Form is emphasized at the expense of content, and nonverbal signals (e.g., tone, pitch, body language) might convey more important messages than do words. Those who are too open and direct might be considered by high-context culture as rude and blunt, and they run the risk of losing face. "Beating around the bush" is often the preferred way of approaching a subject. In contrast, communication in low-context cultures occurs predominantly through open and precise statements in text and speech, to ensure that the listener/reader receives the message exactly as it was conveyed. The mass of information is invested in the explicated code, the meaning depends on the content of the spoken word, and, for that reason, content is emphasized at the expense of context. Directness and being "straight to the point" are considered virtues. The subjects of low-context cultures will shift from information already stated to information about to be given, while high-context cultures will jump back and forth and leave out details, assuming them to be implicit between the two interlocutors.[3]

Perception of time is another parameter for measurement of cultures offered by Hall. On the "cultural time" spectrum, high-context cultures tend to be polychronic ("attend to many things at once"), while low-context cultures are monochronic ("one thing is considered at a time"). "Polychronic time" is not structured and flows boundlessly. Individuals from high-context cultures believe that everything will happen in its time and that "everything is connected to everything else." The subjects of these cultures are cyclic when in the workplace, changing from one activity to another. Their mode of thought grasps a lot of things but does not concentrate on something particular for long. For the subjects of the monochronic culture, time is a fixed, measured resource, divided into concrete intervals. Activities linked to these blocks of time are detailed, planned, and scheduled. Monochronic individuals from low-context cultures do not switch between missions but concentrate on one activity at a time until it is fully accomplished.[4] Works of Charles Cogan, Richard Solomon, Raymond Cohen, Edmund Glenn, and Robert Bathurst demonstrate how the above cultural differences frequently manifest themselves in international security, negotiations, and business, when the interaction occurs between low- and high-context cultures.[5]

The cultural continuums of *context* and *time* usually parallel the dimension of *collectivistic* versus *individualistic* societies, introduced by Geert Hofstede. People with a high-context cultural orientation usually come from hierarchically structured collectivistic societies that prioritize group values over the goals of the individual. People in collectivistic societies tend to be interdependent through a network of deep-rooted relationships. There is an emphasis on conformity and group orientation, and the society as a whole has a strong welfare mentality. Solidarity between people is seen as the ideal and the normative objective. People from low-context cultures usually arrive from somewhat fragmented and highly individualistic societies, which tend to emphasize the goals and independent accomplishments of the individual rather than the group. Low-context culture is usually spread among action, assertiveness, and achievement-oriented societies, where individuals are motivated by self-interest.

Another characteristic tied to this continuum is the *power distance dimension*—the extent to which less powerful members expect and accept unequal power distribution within a culture. Societies with high power distance include many hierarchical levels and an autocratic leadership, expect inequality and power differences, and usually parallel collectivistic high-context cultures. In contrast, low power distance societies are characterized by flat organizational structures, consultative or participatory management style, and an expectation of egalitarianism. In low power distance societies, superiors encourage independence and initiative in subordinates. By contrast, in higher power distance countries a very clear division of power exists, and different responsibilities are assigned to each player. High-context cultures are less prone to delegate authority downward; innovative ideas and planning usually originate at the top. In low-context cultures, on the other hand, responsibility is diffused throughout the system, and innovations are generated from the bottom up.[6]

Cognitive Continuum

Psychologists and anthropologists argue that the sociocultural differences discussed above underlie both chronic as well as temporary variations in cognitive style across cultures.[7] Cognitive style constitutes an individual's preferred collection of strategies to perceive, organize, and process information.[8] These styles vary along the continuum between two diametrically opposed patterns: one grouped under the heading of *holistic-dialectical thought* and the other under the heading of *logical-analytical thought*. Cognitive psychology defines *holistic thought* as a tendency to attend to the context or field as a whole. Sometimes scholars dub it *field-dependent* reasoning. This mode of thought focuses on and assigns causality to the relationship between a focal object and the field, and explains events on the basis of such a relationship. Holistic approaches rely on experience-based and intuitive knowledge rather than on formal logic and are dialectical, meaning that there is an emphasis on change in the context, recognition of contradiction, and search for the synthesis between opposing propositions. *Analytical thought*, sometimes defined as *field-independent*, tends to center attention on the focal object, to detach it from its context, and to assign it to categories. This mode of reasoning uses formal logic to explain and to predict an object's behavior. Analytical thought, similar to what is defined by Thomas Kuhn as "normal science," avoids contradictions and does not suffer from conceptual discontinuities.[9]

Physiologists speculate that the origin of these differences is traceable to markedly different social systems. On the basis of statistically significant evidence, researchers claim that individuals socialized in high-context, collectivistic social structures tend to rely on holistic reasoning. On the other hand, analytical thinking prevails among persons coming from low-context, individualistic societies.[10] Similar to cognitive psychologists, anthropologists argue that conceptual styles are associated with different socioeconomic structures. Holistic thinking exists in cultures where the social order approaches a collectivist pattern, while analytical thinking tends to prevail in societies that have an individualist pattern.[11]

By its nature, the holistic-dialectical approach is better suited for grasping emerging changes in the relationship between focal point and context than is analytical-logical reasoning, which decontextualizes objects. "Analytical attention" to the object encourages its categorization, the application of rules to it, and use of it as a point of reference for causal attribution. "Holistic attention" to the field encourages noticing relationships and change, and prompts causal attribution in terms of the context and distal forces.[12] Although holism lacks the power of precise analysis and systematic classification, it can excel in explaining phenomena by pointing to other events that occurred at the same time in a broader context, even though by analytical logic the events are unconnected.[13] It should be stated, however, that a possible correlation between the sociocultural structure and thought patterns does not mean that an individual from a high-context culture is unable to apply formal logic or that a low-context culture's individual is incapable of making a holistic observation.[14] A specific culture's position on the cognitive-styles continuum is relative.[15]

Some scholars parallel logical-analytical and holistic-dialectical cognitive styles with inductive and deductive thought patterns. Whereas inductive thinking aims to derive theoretical concepts from individual cases, deductive reasoning aims to interpret individual cases within previously derived theoretical concepts. These different thought patterns manifest themselves across different cultures and have a profound impact on argumentation, communication styles, and the way the world is seen and understood. Whereas Anglo-Saxon thought patterns are predominantly inductive, Latin American, Eastern, and Russian thought patterns are predominantly deductive. Thinking within the Aristotelian logical tradition, which is dominant in most Western cultures, may not be understood by people from a culture that emphasizes a more holistic approach to thinking.[16] The impact of cultural environment is clearly demonstrated in the social history of science. Scientists from contrasting cultural backgrounds working on the same questions take different cognitive approaches and reach different conclusions.[17]

Cognitive Styles and the Pass to RMA

How, intentionally or inadvertently, do different militaries organize their thoughts and their subsequent actions when they embark on military innovations? As Thomas Mahnken has shown, military institutions innovate in three distinct, yet often overlapping, phases: speculation, experimentation, and implementation.[18] This book focuses on the intellectual activity that takes place during the speculation phase. In this initial stage, military innovators identify novel ways of solving existing operational problems and of exploiting the potential of an emerging technology. It is also the earliest opportunity for experts to realize revolutionary conceptual and organizational innovations that new technology can put into motion. At the point at which military innovators start to identify novel ways to exploit the potential of technology, epistemology—the ways and means of production of knowledge—comes into play.

According to Richard Hundley, identifying RMA involves a conceptual breakthrough in the way we think about warfare. To recognize it is, in other words, to attend to a change that occurs in military regimes at a current phase.[19] In line with this view, Boot argues, in his comparison of RMAs across human history, that a military revolution, like a scientific one, demands a paradigm shift from one set of assumptions to another.[20] Thus, RMA is a paradigm shift in the nature and conduct of military operations.[21] The paradigm, as we know from Thomas Kuhn, is not only the current theory, but a conceptual set of assumptions about reality that allows one to process data, to elaborate theories, and to solve problems. Paradigmatic change does not take place at the level of mere rational discourse. This shift in perception is not a function of a linear progress moving toward more accurate and complete knowledge, but a radical shift of vision in which a multitude of nonrational factors come into play. When a new paradigm is eventually articulated, it brings fundamental changes in theoretical assumptions and transforms the entire worldview in which it exists.[22]

Paradigm in military affairs is an accepted model that serves as the basic pattern for waging operations.[23] In keeping with this line of thought, one may argue that recognizing paradigmatic shifts in military affairs is attending to a major discontinuity in the relationship between the *object* (technology) and its *context* (organizational structures and methods of force deployment). Technology (the object) is relatively easy to acquire, compared to doctrines and organizational structures (the context), which are far more difficult to conceptualize and implement. However, only when technology is enveloped by a "new theory of war" does a quantum leap into the Revolution in Military Affairs occur.[24]

Since RMA is an intellectual product of military experts, the cultural context of developing military knowledge seems to matter a great deal. The aptitude to envision and to perceive the discontinuity in a military regime is effected, inter alia, by cultural factors. These factors, argues Farrell, significantly shape a national approach to innovations, adaptations, or emulations—three paths to major military change.[25] Ample empirical evidence leaves little doubt that under the impact of cultural factors some conceptual approaches became possible and others less so when the same military technology came into the possession of different nations.[26] One may assume, therefore, that under the impact of sociocultural factors, certain styles of thinking will have a greater aptitude than others to grasp paradigmatic changes in military affairs.

Keeping in mind the dialectical nature of the RMA, and consistent with accepted definitions of cultural-cognitive inclinations described above, an argument can be made that a defense community predisposed to a logical-analytical cognitive style would be less receptive to recognizing the emerging RMA than would be theoreticians that employ holistic-dialectical modes of reasoning. Recognizing paradigmatic shifts in general, and an RMA in particular, involves more deductive than inductive thinking, because it demands elevation of the whole over the parts. Observing the same "object," subjects from cultures predisposed to a deductive pattern of thought may consequently possess more aptitude to recognize paradigmatic shifts in military regimes.

Hundley, in his study of the intellectual roots and epistemic foundations of the RMA, argues that RMAs result from serendipitous conceptual breakthroughs in the minds of military experts. He defines this effect as an "essential creative element in the heart of the RMA processes."[27] *Serendipity effect* is known in psychology and social sciences as an accidental discovery or invention, which occurs as a flash of insight while looking for something else entirely than the focal point of interest.[28] Being accidental and associative, these significant leaps in understanding are not driven by linear formal logic. On the contrary, these breakthroughs seem to be driven by a holistic style of thought, because the revelation originates from the associative observation of objects from the context. A polychronic culture may be inherently more predisposed than a monochronic one to recognize and to understand the interdependence of the objects in the field.[29] This quality also contributes to the ability of holistic-dialectical reasoning to reveal an RMA more quickly than with logical-analytical cognitive style.

Certain individuals and organizations in a given intellectual climate may vary in their interpretations of the details of the national security discourse. However, the main vectors of national thinking style on security affairs remain constant when a defense community in its entirety embarks on military innovation. Empirical evidence shows that military theoreticians from the three countries developed their own peculiar approaches to military futurology and conceptualized future war differently. This implies that distinct techniques for generating knowledge in military affairs do indeed exist in different countries. It follows, then, that understanding the ways of reasoning used by professionals from a given strategic culture can produce a solid hypothesis about their aptitude to recognize RMA. Such a hypothesis might bring us closer to accounting for the divergence between the USSR, the US, and Israel with regard to the RMA. On the basis of analytical models from cultural psychology, the following chapters will describe the cognitive styles of Soviet, American, and Israeli defense experts. It will reveal how, in tandem with other strategic cultural traits, their particular methods of reasoning affected ways in which they thought about the "new theory of victory" and envisioned innovative applications of a new technology.

It is hard to definitively establish the significance of the explanatory contribution of cognitive psychology to security studies literature and to the strategic culture approach. Cultural-cognitive psychology is a young discipline, in the midst of developing its research methodology. Consequently, the demonstration of strong causation has so far dogged attempts by scholars in the field. Indeed, the following chapters will indicate that the cognitive style explanation appears secondary in each account to traditional cultural analysis.

However, "cognitive styles" theory has added value to the strategic culture literature since it contributes to our understanding of the way militaries think about warfare and sheds light on intellectual climates within military organizations. As the framework of analysis discussed in this chapter indicates, the theory of cognitive styles has much to contribute in explaining the sources of disparities in intellectual approaches to military innovations.

2 THE IMPACT OF CULTURAL FACTORS ON THE SOVIET MILITARY-TECHNICAL REVOLUTION

ALTHOUGH the US laid the technological groundwork for the RMA, Soviet military theorists were the first to intellectualize about its long-term consequences. Since the late 1970s, the Soviets were pioneers in championing the argument that the new range of technological innovations constituted a fundamental discontinuity in military affairs. The Soviet military theoreticians dubbed it the *Military-Technical Revolution* (MTR) and produced seminal theoretical works without possessing either weapons or technologies. This chapter traces the impact of the Russian-Soviet strategic culture on the approach by the Soviet military to the MTR.

Following the introduction, the discussion will be divided into three parts. The first part reconstructs the intellectual history of the MTR in the USSR. It discusses the Western doctrinal reaction to the Soviet echelonment technique, describes the conceptual birth of the Soviet MTR, touches on the implications of this innovation for operational-tactical levels of the Soviet military, and analyzes factors that prevented it from implementing its revolutionary theories. The second part concentrates on the general sources of the Russian-Soviet strategic culture. It describes several fundamental societal and cognitive characteristics and the relevant traits of the strategic culture, such as: the "Russian way of war," a role of technology in military tradition; scientific organization of the military thought; a role played by the General Staff; theory development versus implementation and practice. The conclusion integrates the previous two parts and uses the traits of the Soviet strategic culture to account for the conduct of the Soviet defense community. Elaborating on the cultural explanations, it clarifies why the technologically inferior USSR prevailed qualitatively over the West in the conceptualization of the emerging military-technological realities, but never implemented them.

PART ONE: SOVIET MILITARY-TECHNICAL REVOLUTION

Soviet Echelonment Doctrine and Origination
of ALB and FOFA

Soviet force development policy addressed two basic issues during the 1960s. First, because the mobility of the enemy's nuclear weapons allowed the massing of firepower over great distances, friendly forces had to be dispersed to avoid an enemy nuclear attack. Second, the same maneuver forces had to mass to break through the breaches in the enemy's defense. The Soviets found the solution to the problem of these mutually conflicting requirements to *disperse* and to *mass* at once. To decrease vulnerability to nuclear attack, forces would be deployed in several echelons in the Soviet rear. When the offensive got under way, they would all concentrate at the line of contact with the enemy, too close to enable NATO forces to use tactical nuclear weapons. Where enemy defenses had been destroyed by Soviet nuclear strikes, the first-echelon forces would open breaches through which the echelons that followed could move rapidly, maneuvering and striking in depth. This strategy was expected to collapse the NATO defensive front.[1]

In the mid-1970s NATO and the US Army became cognizant of the Soviet technique of echelonment and realized that in their current state, their defenses could not stand up to Soviet conventional superiority. The West found a remedy in emerging technologies. Since the mid-1970s, highly advanced technological achievements, particularly in the field of microprocessors, computers, lasers, and electronics, had enabled the production of so-called smart weapons—an assortment of conventional munitions that were precision-guided to targets—even in stand-off ranges. Developments in weapons technology and evolution of thought about future war in Europe led to similar warfare missions in the US and NATO: to *strike deep* against enemy offensive follow-on forces.

The US response was a *deep attack* doctrine known as *US Field Manual 100–5* or Air-Land Battle (ALB). It rested on the premise that follow-on Soviet echelons had to be neutralized before engaging with the NATO forces. Attacking the second echelons and destroying them determined the US tactics and weapons development. The ALB envisioned attacking the Soviet second echelons with both stand-off precision fire and ground offensive operations.

Deep attack required improved air support and more accurate long-range-fire capabilities.[2] The "father" of the ALB, Gen. Donn Starry, believed that deployed conventional forces could continue to fight the advancing Soviet first echelon, and in the meantime, before the second echelon had time to regroup after its initial deep-strike bombardment, release authority for battlefield nuclear weapons could be obtained from US political leaders.[3] Similarly, the impetus for NATO's Follow-on Forces Attack (FOFA) evolved from the same concern and was fueled by the development of emerging technologies. During the 1970s, steps were taken to reduce the ratio of enemy forces arriving at NATO's defensive positions by using conventional munitions only. FOFA, like its US counterpart, was designed to attack the enemy as far into the rear as the target acquisition and conventional weapons systems permitted. Final FOFA proposals were submitted to NATO's Military Committee in October 1981. The ALB doctrine was formally released a year later.[4]

Although differences existed between the two doctrines, the Soviets never placed much emphasis on the distinctions between the offensive posture of the ALB and the defensive one of FOFA.[5] The Soviet General Staff (GS) anticipated that the combat outcome of both would be essentially the same—attacks on Warsaw Pact (WP) "operational" depths that would result in disorganization and distraction of its second echelons.[6] Warsaw Pact defense ministers saw developments in conventional armaments in the early 1980s as even more ominous than the strategic change wrought by nuclear weapons developments.[7] ALB and FOFA that encapsulated these capabilities would have been extremely threatening to Soviet defensive and counteroffensive operations.[8] Between the late 1970s and early 1980s, the conceptual task of Soviet military operations crystallized, focusing on how to neutralize the ALB and FOFA without using nuclear weapons that would escalate the conflict into global war.

The Roots of the Soviet Military-Technical Revolution

MacGregor Knox and Williamson Murray trace the conceptual roots of the current Western notion of the RMA to the Soviet writings on the Military-Technical Revolution.[9] Indeed, Soviet theoretical variations on the MTR originated during the mid-1970s and early 1980s. They were an outgrowth of two interrelated professional discussions that took place in Soviet military circles. The first of these discussions was disconnected from the debate over Western military doctrines and started around the mid-1970s, before the seeds of Western military innovations emerged. In keeping with Soviet military theory, it diagnosed a new Military-Technical Revolution and concentrated on the fundamental, futuristic inquiry into the impact of scientific progress on the methods of future military operations. The second discussion focused on military-technological remedies for recent Western doctrinal innovations. It generated an enormous amount of insight that greatly influenced the formulation of the MTR.

Official Russian military history treats this period as a doctrinal competition between NATO and the Warsaw Pact, with an enormous degree of mutual emulation in developing new types of concepts. However, it would be an oversimplification to assert that the MTR concept merely constituted a Soviet response to threatening Western doctrines.[10] While posing the doctrinal antipode to Western "deep striking" capabilities, it went far beyond any particular doctrinal countermeasure and offered a new, coherent theory regarding the future battlefield under the impact of scientific-technological progress. Each of the Soviet discussions constituted a necessary but insufficient condition for formulation of the final version of the MTR concept. Both discussions were symbiotic and complementary. Only the merging of the two succeeded in producing a coherent military theory, which, while relying in part on US and NATO military-technological achievements, was nonetheless highly original.

Soviet MTR ideas diverged from Western Active Defense doctrines, which at that time were far from their final formulation.[11] From the mid-1970s, the Soviets began to engage in theorizing about the third Revolution in Military Affairs—fundamental changes taking place in the nature of warfare under the impact of the new technologies.[12] By the early 1970s, Soviet military theoreticians and forecasters had identified two periods of fundamental change during the twentieth century. The first Revolution in Military Affairs was prompted by mechanization of air and ground warfare and culminated in the appearance of deep operations in the 1920s and strategic bombing in World War II. A second Military-Technical Revolution of the early 1950s was stimulated by the development of nuclear weapons and missile technology.[13]

In the mid-1970s, Soviet military forecasters declared that the highlight of the current phase of military development was the unprecedented emergence of qualitatively new technologies and equipment. Military applications of microelectronics, laser, kinetic energy, radio frequencies, electro-optic, electromagnetic pulse, remote control, and particle beam technologies figured prominently in Soviet professional discussions. Particular attention was devoted to automated decision support systems, telecommunications, and enhancing accuracy, range, and lethality of stand-off and direct-attack precision-guided munitions. The Soviets believed that the emerging technologies would potentially extend the depths to which future systems—both sensor technologies and means of fire—would operate.[14] They believed that the current stage was similar to the early 1920s and early 1950s, when the impact of qualitatively new weapons on operational concepts and force structures transformed the nature of war. Once the consensus about the arrival of the third MTR was announced, military theoreticians were expected to conceptualize the consequences of the change in military regime. The Soviet theoreticians stressed that the future introduction of these new means of combat into the tactical, operational, and strategic levels had to be accompanied by the development of a new concept of operations.[15]

These initial insights were transformed into unprecedented conceptual activity, which gathered momentum under Marshal Ogarkov, chief of the GS since 1977. Following his promotion, a number of senior officers, noted for their writings about technologies and future warfare, were also placed in key positions.[16] In Ogarkov's view, a revolutionary change was under way, due to the qualitative improvement of conventional weaponry. In several professional periodicals and books, Ogarkov continuously utilized the term *Military-Technical Revolution*, indicating that the latest technologies, which made it possible to "see and to strike deep" in the future battlefield, and the organizational changes that would accommodate these capabilities, would not constitute a phase in a process of evolutionary adaptation, but a genuine discontinuity in military affairs. The impact of the "scientific-technical revolution" required exploitation of emerging technologies to invent innovative means of conducting operations and to adjust force buildup and structure in each military service.[17]

Other leading Soviet military theoreticians echoed Ogarkov's premises of the emerging MTR. They attributed the change to a "qualitative leap" in the modernization of the means of armed conflict, resulting from the latest developments in science and technology, and first and foremost as applied to conventional "high-precision" means of warfare. They considered the appearance of these new weapons a turning point in the development of military art.[18] Particularly, the MTR was expected to enable a greater degree of control from the center, and to increase the likelihood of the potential attacks on troops' command and control facilities. Equal importance was attributed to future automated information processing, which would potentially compress the "planning to decision" and "detection to destruction" cycles.[19] Long-range and highly accurate fire systems in tandem with expanded sensor and target acquisition capabilities would increase attrition rates against fixed and mobile targets.[20] Explosive substances of enhanced power would expand the zone of destruction.[21] Surprise, coordination of high-tempo operations, and intensive fire support were emphasized as well. The future battlefield was seen as increasingly complex, with different kinds of forces participating simultaneously in combined arms theater operation.[22]

The MTR discussion coevolved with the Soviet reassessment of the nuclear–conventional balance. Historically, Soviet views underwent a triple transformation during the Cold War. In the 1950s–1960s, the "next war" was envisioned as an unrestrained missile confrontation—nuclear from the outset. The views of the early 1970s anticipated a gradual passage from an initially conventional conflict to total nuclear war. In the late 1970s, the Soviets were reckoning the probability of a dual nuclear and conventional conflict.[23] Beginning in the 1980s, though, they became skeptical about the potential for an outbreak of the former. Moscow argued that within the context of the emerging revolution in military technology, nuclear weapons would continue to play an important but diminishing role in future war, which could well maintain its conventional character throughout.[24] By the early 1980s, the Soviet military recognized the inability of either side to "decapitate" its enemy's nuclear potential from the first strike. The catastrophic consequences of a nuclear conflict, combined with the development of high-accuracy conventional weapons, increased the probability that the next war would be an all-out non-nuclear one. This in turn boosted the development of the new theory of conventional war.[25]

The Soviets vigorously discussed the application of innovative technologies in producing a new generation of conventional armaments that would exceed their predecessors in range, destructiveness, reliability, speed of delivery, and accuracy.[26] The operational effects that could be produced by emerging conventional systems were on a par with nuclear measures. Precision-guided strikes in combination with timely detection of targets approached the effectiveness of tactical nuclear weapons. The tempo and the scale of the conventional operations started to resemble those of the nuclear warfare scenarios. More and more missions and tasks formerly perceived as solely nuclear were expected to become conventional.[27] Marshal Ogarkov called for greater attention to preparing for a war that would employ solely conventional means of destruction.[28] During the 1978 Warsaw Pact exercise, the Soviet counterattack was initially waged with conventional weapons.[29] The possibility of a large-scale conventional war was raised for the first time during the "Iug" exercise in 1980 and during the "Zapad" strategic maneuvers in 1981. The 1982 Warsaw Pact exercise held in Czechoslovakia responded to the US strategy to win a war in Europe by conventional means only. In 1983 the GS formulated its first vision of waging future war through conventional weapons alone.[30] Thus, at the level of grand strategy, the MTR made nuclear war a less desirable option in the eyes of Soviet strategists and shifted the equilibrium toward conventional confrontation. The once-nuclear arms race was transformed into one of technological conventional force multipliers.[31]

The vision of strategic offensive changed gradually in the early 1980s, under the influence of the MTR discussions, which rediscovered strategic defense in Soviet doctrine.[32] Analyzing the depths to which the high-accuracy systems were capable of operating, the Soviets declared that the border that had divided combat into offensive and defensive was increasingly being erased, since these two forms of conducting war were using the same weapons to achieve their operational goals. The MTR made it almost impossible to distinguish between the *defensive counteroffensive* and *pure offensive attack*.[33] The Soviets realized that the choice of the timing of the encounter had ceased to be exclusively in the hands of the attacker, thus rendering the defense far more "active." The defender's ability to reach the enemy at distant pre-battle positions or on march routes meant that he no longer had to wait for the attacker to strike and could make decisions about initiating battle. It was more effective under these circumstances for the defender, rather than retaining territory, to defeat enemy forces by means of stand-off deep strikes. Equipped with a deep-striking precision-guided arsenal, a defender's strikes would not be limited to the traditional tactical zone but would achieve a range across the operational depth of the enemy's deployment.[34] With these new capabilities, the Soviets believed, a defender seemed capable of successfully thwarting an attacker's preparations and launching a counteroffensive under favorable conditions.[35] The MTR made it possible to inflict heavy losses on the enemy whether fighting from an offensive or a defensive posture.

During the same period, the Soviets turned to their past experience for developing new conventional doctrines.[36] Elaborating on the lessons of the Great Patriotic War, Soviet theoreticians were concerned that Soviet military art had become preoccupied with offensive operations at the expense of a more balanced offense–defense mix.[37] The availability of "deep" defensive strikes through the entire depth of the enemy's deployment could buy time for completing mobilization and improving the conditions for an imminent counter-offensive.[38] From 1984 onward, massive theater maneuvers that exercised full-scale strategic defensive operations took place in various military districts of the USSR. The buildup of forces reflected the defensive climate of Soviet strategic thinking.[39] Even before Gorbachev's defensive civilian reforms, and purely out of MTR considerations, the defensive posture of operations began to gather momentum.[40]

Reflections of Western Doctrines in Soviet Professional Literature

Although Soviet military forecasters had developed a significant range of their own vision even before the introduction of deep-striking concepts in the West, beginning in the late 1970s, they inquired intensively into Western doctrinal innovations. The professional military periodicals indicate that the Soviets had identified Western technological developments in the field of conventional warfare almost immediately after the first discussions were held in the US in the mid-1970s on deep-strike capabilities. It happened before the official US and NATO shifts had been made in favor of ALB and FOFA in the early 1980s.[41] The trend in the West towards an emphasis on new conventional weapons was carefully monitored. In 1978, the military intelligence warned that NATO's technological surprise moment might be coming. A special emphasis was given to the use of innovative deep-strike features against the second echelon of the advancing adversary.[42] The Soviets saw Western development of the conventional technologies as a confirmation of their initial assumptions about the MTR and exploited them as an auxiliary frame of reference to consider how emerging technologies might be used in future war.

From 1978 onward, *Zarubezhnoe voennoe obozrenie* (*ZVO*) and *Voennyi vestnik* (*VV*) published an unprecedented number of articles that recognized that emerging technologies were combining precision target-location capability[43] with increased distance striking capacity from the land[44] and air,[45] and linking them via automatic command and control equipment.[46] For the next three years, *ZVO* focused intensively on emerging Western capabilities and by early 1980 reflected a clear realization that such a combination would allow a *deep stand-off striking* capability, either offensive or defensive, enabling adversary formations to hit with forces located far behind the zone of immediate contact.[47] By 1980 and onward, in the classified issues of the Soviet General Staff's *Voennaia Mysl'* (*VM*), the flood of articles on Western military-technological innovations was accompanied by profound discussions of US and NATO military doctrines based on the new *stand-off precision-guided munitions*. Analyses concentrated on the emerging ALB and on the possible adaptations of such deep-fighting concepts by NATO forces.[48] Ogarkov and other senior military figures expressed their concerns that Western advances in military technologies could offset Soviet advantages in the conventional balance.[49]

By debating these conceptual innovations and by analyzing NATO exercises, the Soviets reconstructed the operational logic of Western doctrine. Among the conclusions, three were especially stressed by the Soviets: deep-strike precision-guided and reconnaissance capabilities were deemed capable of destroying the second echelons of the potential enemy; the previously limited non-nuclear stage in any given conflict had increased considerably; and an increase had occurred in the volume of tasks resolved by troops utilizing only conventional weapons.[50] The Soviets argued that the emerging ability to conduct massive and precise conventional strikes into the entire depth of the operational rear could, at the very least, disrupt the successful implementation of assigned missions and, at worst, have profound strategic implications for the entire front of operations.[51]

An analysis of Soviet military publications indicates an exceptionally sophisticated level of professional understanding by the Warsaw Pact officers of Western military-technological innovations. What is even more striking is that the Soviets were able to place these innovations in a much deeper and broader context, reflecting a far more profound grasp of these developments' implications than the West itself possessed. For Western military leadership, the massive equipping of forces with new means of combat in the framework of ALB and FOFA represented little more than the development of a new form of combat action against the enemy's second echelons on the Central Front. However, in Soviet eyes, the ability "to see and to strike" through the entire depth, precisely and simultaneously, was treated as a yawning discontinuity that had significant ramifications in terms of the methods of employing corps and armies, and that shaped the nature of war in a revolutionary way. The Soviets saw ALB and FOFA as much more than simply a doctrinal update or an operational threat and sought broader theoretical frames of analysis in order to describe these developments.[52] The insights acquired through inquiring into the operational concepts of the West and seeking a countermeasure provided the Soviets with a frame of reference and intellectual fuel for the MTR concepts.

Learning from Western innovation constituted the multiplier for previous and subsequent Soviet variations on the topic.

The Implications of the MTR for Soviet Military Power

Describing the dialectical relationship between the scientific-technical revolution and military science, the Soviets emphasized the primacy of the latter over the former.[53] Soviet sources assumed that equipping the military with PGMs and new means of reconnaissance and control would occur at an equal rate on both sides; thus, superiority would be gained by whichever side realized the concept more rapidly and broadly.[54]

The Soviets argued that the emerging forms of weapons made war in the MTR era extremely dynamic: forces could now attack with a smaller density of personnel and equipment, yet at considerably greater depth and with greater momentum.[55] They deduced that given the expanded scope and unprecedented tempo of the modern battlefield, and especially in light of the shifts in nuclear–conventional and offense–defense balances, previous forms of employment of military forces had ceased to correspond to existing conditions. Ogarkov argued that mere modifications and adjustments could no longer produce the desired results, and that an elaboration of new forms of combat operations was urgently needed.[56] The theoretical debate about the impact of the MTR on the vision of the future battlefield resulted in the development of three interrelated concepts: reviving of the "deep operations battle"; the reconnaissance-strike (RUK) and fire (ROK) complexes; and the concept of operational maneuvering groups (OMG).

Soviet military science was the science of future war that operated, however, within strict confines of the past. Traditionally, the assessment of the current generation of new doctrinal models leaned heavily on historical operational analyses. Analytical study of military history was considered essential for predicting future developments in warfare.[57] Thus, the fresh vision posited by the GS led many to turn to the experiences of WWII and to the old doctrines in search of new operational forms. Since the MTR elaborated on the notion of applying the new technology in conducting operations in "depth," the Soviets had no need to reinvent the wheel; they had merely to revive the Deep Operations theory of the 1930s. When the new technologies arrived in the mid-1970s, the Soviets already possessed the conceptual consciousness, the scientific framework, and the advanced system of military terminology for bridging the gap between the abstract (operational concepts) and the mechanical (high technology) in the extended battlefield. Professional military publications clearly indicate that Soviet theoreticians turned to Tukhachevsky's theory of *deep battle* of the early 1930s, applying its essence to contemporary and future operations. The authors pointed out the relevance and the importance of emulating this early concept for modern operations.[58] In 1976 Marshal Ogarkov himself contributed a historical article to the Soviet military encyclopedia on the concept of operations in depth. He maintained that the general principles of the original Deep Operations theory had not lost their significance.[59]

Combining this early concept with the new, more advanced technology potentially meant significantly increasing the intensity of the strike to the entire depth of the enemy, penetrating its defense, and establishing a rapid pace for developing swift "operational success." The Soviets stressed the need for better command, control, and methods of targeting, and they noted the increasing importance of the deep rapid maneuver under long-range precision fire.[60] In Ogarkov's view, the "theater-strategic operation," that is, orchestrating a number of armies and fronts in a simultaneous single action, had to be instituted as the principal method of waging a campaign. To execute such an operation successfully, the military had to be able to achieve *deep targeting* within the theater, so that ground forces would be able to rely on theater fire support to the full depth of their deployment.[61] The Soviets argued that the modern air-land operation would be three-dimensional combined arms battles fought simultaneously on the forward edge of contact and in the depth, on the ground and in the air.[62] In theory, the concept of the modern MTR era operations coalesced into the notion of simultaneous strike against the entire depth of the enemy's operational structure by stand-off precision weapons.[63] The innovative means for implementing this concept were primarily twofold: reconnaissance-strike/fire complexes and the operational maneuvering group.

The chain of developments that led to the birth of RUK and ROK was as follows. The *deep-strike* PGMs incorporated in ALB and FOFA stimulated the Soviet military to create its own concepts and combat architecture to invalidate Western deep-strike measures and to improve capabilities against the enemy's depth.[64] Soviet theorists realized that given the tendency toward greater mobility and deception, the time available between acquisition of the target and the ability to destroy it would be limited. This demanded the establishment of a complex combat architecture that would consolidate the means of reconnaissance with high-precision, fire-destruction elements, linked through command and control channels. The quintessence of that ability was labeled *reconnaissance-strike* and *reconnaissance fire complexes,* which constituted an all-encompassing implementation of the Soviet MTR. This "system of systems," which was to consist of an integrated triad of ground, air, and space reconnaissance, surveillance, and target acquisition assets; deep-strike weaponry; and advanced command and control, was designed to conduct the war over much greater distances and with greater precision, co-ordination, and tempo than ever before.[65] The first, although somewhat obscure, mention of the *reconnaissance fire complex* appeared as early as 1979,[66] and by 1982 the WP officers were already acquainted with ROK fire coordination.[67] Laborious discussions on the definition of RUK and ROK emerged in *ZVO* and *VV* from 1983, where the concept was attributed to the US military,[68] not unlike the Soviet military dictionary, which defines it as a "foreign term."[69]

Though the West was the first to develop all the military capabilities attributed to this concept, US defense specialists who analyzed the MTR referred to RUK and ROK as an exclusively Soviet innovation with no analogies in the Western military vocabulary.[70] According to American sources, the Soviets, while analyzing the tests of systems such as *Assault-Breaker* in the late 1970s, gave the US military far more credit than it deserved for developing the *reconnaissance-strike complex.*[71] The technologically inferior Soviets relied successfully on the Western scientific advantage and exploited it as a starting point for their conceptualizations. The Americans treated the Soviet RUK and ROK concept as revolutionary, studied it intensively, and later emulated parts of its rationale.[72] Though the US planners devoted greater resources than did the Soviets to the conventional weaponry, the latter paid greater attention to the ideas of the "integrated battlefield." Soviet observers actually provided the initial historical argument in the 1980s that US forces were exhibiting revolutionary capabilities.[73]

The concept of the operational maneuvering group was, at least in theory, a maneuvering component of the MTR that "leaned into" long-range supporting fire and intelligence of ROK and RUK.[74] OMG was a reworked version of the mobile group concept from the Great Patriotic War, when autonomous armor formations, using stealth and mobility, infiltrated the enemy's operational rear and, using shock and firepower, created command and control chaos from within. Modified in light of the deep precision strike and enhanced intelligence capabilities, the essence of the OMG adhered to the principles laid down in Tukhachevsky's original theory. The OMG concept committed part of the force across the front much earlier and deeper, to avoid an ALB and FOFA attack, thus carrying out a Soviet preventive blow into NATO's rear. Swift infiltration of a group of armored divisions along several axes would create a deep and dynamic center of gravity in NATO's rear. It would turn over the defense, create operational shock, paralyze the enemy's ability to react, and result in operational disorganization.[75]

In theory, the OMG consisted of a reinforced combined-arms armor division that operated in conjunction with fixed- and rotary-wing air support. At least two OMGs would operate autonomously on different axes on the fragmented and nonlinear deep battlefield in the mode of encounter engagements intended against PGMs, command and control, intelligence capabilities, or the tactical nuclear weapons of the enemy.[76] The OMG concept was inaugurated during the *Zapad-81* exercise. The maneuvers practiced deep conventional thrusts of the Pact armor groups into NATO territory to disrupt its military infrastructure. The WP *Soiuz-83* exercise practiced extensively the deployment of the OMGs in the Western Theater of Military Operations.[77] At the later stage of concept development, the coordination between ROK/RUK and OMG resulted in their eventual organic unification under the term the Reconnaissance-Fire Group. In theory, intelligence assets, stand-off fire capabilities, and maneuvering elements of the extended battlefield were to be orchestrated as one integrated whole.[78]

In the late 1970s and early 1980s, the GS worked on the seminal classified publication *The Fundaments of Preparation and Execution of Operation in the Soviet Military Forces.* This five-volume work was drafted by Gen. A. D. Danilevitch, the deputy of the GS Operational Directorate for military theory and a close associate of Ogarkov. The publication incorporated insights and remarks from most of the Soviet senior leadership and constituted the general guidance for waging all spectrums of operations under the MTR. From 1980, when the critical mass of concepts was developed, Ogarkov initiated several war games, exercises, and maneuvers to test and to experiment with the innovative visions presented in the volumes.[79] The MTR visions were tested during the exercises *Iug-80* and *Zapad-81*, and after learning the initial lessons, its quintessence was embodied in *Zapad-83* maneuvers. The goal of the maneuvers was to train forces to operate in war waged by PGMs from both sides. The specific tasks included the defense against the enemy's PGMs, counterattacks of the enemy's ROKs and RUKs, and training command and control of the

Soviet ROKs and RUKs. Special attention was paid to the use of ROKs and RUKs in the forward engagement and their interoperability with the OMGs. The *Zapad-84* maneuvers were a variation on the previous year's theme. They carried out training in combat concepts for reconnaissance-strike systems on the operational-tactical and the strategic levels. According to Gen. V. I. Varennikov, then chief of the GS Operations Directorate, the maneuvers confirmed basic Soviet assumptions about the nature of MTR era warfare and produced enormous insights for improving operational concepts and force organization. On the operational-tactical level, MTR provided the ability to conduct simultaneous action, by fire and maneuver, against the enemy to the entire depth of their operational formation.[80]

Why Did the Soviets Fail in the Implementation of the MTR?

The Soviet military championed the conceptual breakthrough in military affairs but never fully implemented it. The operational execution of MTR ideas and massive fielding of MTR weaponry was beyond the political, economic, and cultural capacity of the Soviet state.

The Soviet version of the MTR, which became associated with an unpopular chief of GS, had not accelerated, lacking bureaucratic support. Soon after his nomination, Marshal Ogarkov, the intellectual engine of the Soviet MTR, fell into professional disfavor. Although serving under Minister of Defense Dmitry Ustinov, Ogarkov overcame his superior professionally and intellectually and often outmaneuvered the minister in his contacts with senior Soviet bureaucracy. Ustinov, however, tried to subordinate the GS to the Ministry of Defense by any means. Ogarkov's relationship with Ustinov deteriorated following his opposition as a chief of GS to intervention in Afghanistan. The split grew deeper with additional disagreements over questions on military and procurement policies and on the nature of the Soviet military presence in the Third World. In this atmosphere Ogarkov's proposals to reform the military in accordance with MTR principles fell on deaf ears and could not muster necessary support from the Kremlin, the Foreign Ministry, or the KGB. The rift between Ustinov and Ogarkov was probably the deepest between a defense minister and a GS chief in the history of the Soviet Union. This dynamic surrounding Ogarkov ultimately led to his ouster as chief of the GS. Ogarkov was demoted in September 1984 and moved to the position of CINC Western Theater of Operations, stationed in Poland.[81] Ogarkov was held in high professional esteem in the WP. He continued to implement his ideas; however, his authority and influence were extremely limited. He was removed from his post in 1988.[82]

In a broader sense, from 1985 the general political climate of perestroika in the USSR was not suited to military innovation. Although Ogarkov's MTR concepts were still afloat, the massive defensive reform overshadowed all previous ideas and innovations. By the end of the Brezhnev era, the GS suggested that the US might surpass the Soviet Union, thanks to the American Military-Technical Revolution. It argued for the necessity of increasing the Soviet military budget, shifting decision making on industrial priorities back to military control, and reorienting procurement toward the kind of investments to which the West had shown its commitment. This advocacy for major budget increases was unthinkable during perestroika; the political leadership was reluctant to allocate the necessary resources for implementation of the MTR, and civilian experts, who began to take a leading role, advanced the argument of "defensive sufficiency." A political course contradicted the MTR vision and hindered support for building and fielding MTR-type weaponry.[83]

In addition to political obstacles, implementation of the MTR was hampered by economic factors. The USSR and, later, the Russian Federation never possessed the necessary economic capacity to embark on such an ambitious military transformation. In one of his publications of the mid-1980s, Ogarkov stated that the new weapons and military technologies were not yet available in sufficient quantities to initiate a new stage of Soviet force development. He expressed his concern that, given these limitations, the existing methods would be preserved with no revolutionary changes in the force structure and concept of operations.[84] Ogarkov's pessimistic assessment proved prophetic. The Soviet Union never succeeded in creating the economic preconditions for implementing the MTR. Before the era of Gorbachev, it had been possible to manage the economy authoritatively and to reorient production and acquisition through a decision dictated from above. But back in the Brezhnev and Andropov eras, there was no bureaucratic willingness to do so, while under Gorbachev, major economic collapse made such a reorientation infeasible and undesirable for political reasons. Even in post-Gorbachev Russia, a deteriorating and outdated Russian military that undertook several defense reforms did not enjoy the economic and material preconditions for a genuine military transformation. Its military science and theory, however, are advanced, futuristic, creative, and sophisticated as usual.[85]

The Soviet campaign in Afghanistan reflects considerable Soviet technological inferiority and a fundamental inability to implement the MTR vision for material reasons. Although the Soviet military realized the potential embodied in reconnaissance, command and control, and fire, it was not able to link these elements together across the Afghan battlefields. Some operations were designed and executed along these principles, but they were marginal and localized.[86] The Soviets started to formulate new concepts for waging nonlinear warfare suited to operating in battlefields dominated by PGMs, to turn away from traditional formations, to redefine echelonment concepts, and to reorganize its forces in a more flexible way. However, on balance, during the campaign in Afghanistan the Soviet military was increasingly unable to cope with the military-technological realities brought about by the MTR.[87] The MTR vision of war was oriented toward large-scale, conventional, air-land, combined arms battle. Afghanistan, however, obliged the Soviets to wage a counterinsurgency campaign.[88] Moreover, innovative concepts such as ROK/RUK and OMGs were massively exercised for the first time only in 1983. They had just begun to emerge and to mature when the Afghan war was approaching its end. Ogarkov, as an opponent of the Afghan campaign in general, was probably not overly enthusiastic about using Afghanistan as a testing ground for innovative concepts.[89]

PART TWO: RUSSIAN-SOVIET STRATEGIC CULTURE

Cultural Characteristics and Cognitive Style

The normative image of Russian culture is collectivistic.[90] The sense of commonness with *narod* (the people) is a powerful and necessary concept in Russian culture. Communalism was not an invention of the Soviet era, but an innate characteristic of Russian society. For generations collective will took priority over individual needs and rights in Russia, and mutual dependence was a unifying factor. Russian-Soviet collectivistic society was organized hierarchically, and the collective mentality emphasized group-centered relations.[91] Individuals in collectivistic Russian and Soviet cultures lived in a complex and interdependent social world with prescribed roles. The scholars who observed Russian and Soviet social interaction and communication styles defined it as a high-context culture. The products of this cultural environment predominantly express themselves in indirect, cyclical, understated, and vague language in text and speech, relying on the listener's or reader's ability to grasp the meaning from the context. Form, ceremony, and the expression of respect are emphasized at the expense of content; nonverbal signals might be more important than words. In negotiations the message is permitted to evolve without referring to the problem directly.[92]

Using Hall's terminology, scholars define Russians as a polychronic culture. Owing to its agricultural, seasonal heritage, Russian time is cyclical and unpunctuated. People live and operate in several time frames at once and are preoccupied with pursuing different tasks simultaneously in an interconnected way. The subjects of this culture do not see time as a finite resource, structured in a sequential and linear fashion. They usually do not tightly schedule themselves in concrete time frames, and while time is money for practical Americans, and punctuality a virtue, being or doing something on time is relatively alien to Russian culture. The venerated virtue in Russian tradition is not punctuality but patience.[93] Polychronic interconnectivity brings inconsistency and contradictions. Dubbed sometimes as the "puzzling Russian soul" or "the ambivalence of the Russian psyche," this cultural quality is widely described in historical, sociological, and psychological literature.[94] Soviet foreign and defense policy, in rhetoric and in action, often expressed a puzzling combination of contradictory attitudes. In frames of the security context, scholars argue, in Russian strategic mentality "both an inferiority complex and a superiority complex can be simultaneously on display: defensiveness bordering on paranoia, on the one hand, combined with assertiveness bordering on pugnacity, on the other." The Soviet mentality saw no contradiction in combining a very aggressive military strategy with pursuit of peaceful political aims and accepted this ambiguity without intellectual embarrassment.[95]

Similar to other cases in which collectivist, high-context, polychronic cultures exhibit a stronger degree of cognitive field dependency, Russian-Soviet society is usually defined by scholars as one with a strong inclination toward "holistic dialectical" types of thought. Inferential processes of individuals from a Russian-Soviet sociocultural environment have a tendency to attend to the context or field as a whole, to focus on and to assign causality to the relationships between a focal object and the field, and to explain and to predict events on the basis of such relationships.[96] Since Russia is a being, not a doing, culture (in contrast to American mental formations, which favor procedural knowledge that focuses on how to get things done), Russians tend to be more contemplative and oriented toward declarative knowledge that consists of a description of the world. The idea of "natural happening" and "occurrence" is more acceptable to the Russian intellectual tradition than to the American one.[97] Cultural psychologists have observed that Russians see the value of understanding an issue as more important than action.[98] Victor Sergeev, an author of numerous works on Soviet cognitive style in strategic affairs, argued that Soviet holistic thought orientation toward declarative knowledge often neglected the problem of how objects were really functioning, and concentrated instead on the "plan of their construction," that is, how different parts of the given object are related to one another.[99]

These conclusions of cultural psychologists are supported by evidence from the history of ideas and science in Russia.[100] Isaiah Berlin, reflecting on Soviet and Russian culture, argued that the state of mind of virtually all Russian intellectuals was dominated by the belief that all problems and issues are interconnected in a single synthetic system. The search for this system is emblematic of the Russian intellectual approach and is spread among all religious, secular, literary, and scientific circles.[101] For centuries intellectual circles in Russia cultivated a concentration on holistic theories as the most appropriate goal for scientific inquiry.[102] This belief produced a normative demand that science be multidisciplinary and that it generate a synthesis of all the possible subjects.[103] An opposite approach was regarded as a radical fallacy of thinking.[104] This tradition of seeking the integrated whole survived throughout the whole of Russian intellectual history.[105] Decades before works on cybernetics[106] and general systems theory,[107] Russian scientists introduced a unifying discipline that sought the universal principles that underlay all the interacting systems.[108] The majority of Russian and Soviet natural sciences inclined toward a systemic and synthetic approach.[109] This holistic-dialectical tradition was well suited to totalitarian social organization.[110]

The Russian approach to scientific inquiry is symptomatic of the mentality that informs Soviet thinking in military affairs. Donnelly argues that there is an innate Russian cultural ability to see the big picture, with an evident impact of this grasp on theory making, military planning, and procurement.[111] Leites and Bathurst agree that in their interpretations of the world outside, the Soviets regarded unrelated details as symptomatic of major trends and often perceived connections between events where Americans saw none. The Soviet military theoreticians, in thinking about war, interpreted each event as a sign signaling a change in the whole and, in keeping with Russian polychronic, holistic culture, saw problems as interrelated and requiring a generalized frame of reference to determine their meanings.[112] According to Naveh, at the center of the Soviet holistic approach for studying the nature of war stood the idea of envisioning the enemy as a system. Thus, the operational logic formulated in frames of this approach dictated neutralizing an enemy system's ability to attain one's goals. This perception of reality produces an abstract yet logical approach for planning operations.[113] Soviet military thought was multidimensional, counting intangible assets such as coherence and active measures as an integral part of the correlation of forces. As such it was more holistic than the Western one.[114] The cognitive maps of the Soviet and American leaders during Cold War strategic interactions demonstrate that the Soviets were driven by a holistic-dialectical thinking style with a strong predisposition toward declarative knowledge.[115] Russian military thought still adheres to the synthetic and holistic method of analysis, which guides it in developing concrete military concepts and procurement policies.[116]

Material and Morale Factors in the Russian Culture of War

The unpredictability of life and extreme physical conditions were the outstanding feature of Russian existence for generations. Life proceeded in the shadow of an approaching geographic, economic, or military catastrophe, and it was not success and comfort, but survival, that was the primary objective for Russians.[117] Throughout their history, Russians exhibited legendary patience under the most brutal circumstances. Shortcomings, suffering, and the need to endure were seen as a part of the normative reality.[118] In contrast to the American pioneering ethos of conquering the wilderness, the Russian tradition saw the domination of nature as an insurmountable task, and focused on overcoming internal rather than external obstacles.[119] Personal triumph over insufferable circumstances encouraged values of self-restraint and moral and physical fortitude.[120] Endurance and endless stamina, the ability to overcome virtually unbearable conditions both in war and peace, are regarded in the Russian narrative as supreme values.[121] As sociological polls indicate, even today the average Russian respondent places an overwhelming emphasis on *endurance* and *tolerance* in the face of hardship as the core characteristic of his approach to life.[122]

This cultural narrative aligns with the broader Russian military tradition. Deliberately or inadvertently, the Russian culture of war has facilitated maximal exploitation of this fundamental societal characteristic.[123] Throughout its history Russia was confronted by adversaries whose societies, technologies, and economies were superior to its own. From the earliest times Russian rulers counterbalanced this persistent inferiority with the one resource in which the country was rich: the power of human masses.[124] From the time of Peter the Great, the Russian high command turned the passive masses into a disciplined and effective army and navy.[125] For centuries Russian rulers made up for material and technological inferiority by capitalizing on the "backward" social institution of serfdom, which enabled them to generate enormous military power.[126] Strategy, generalship, and operational art relied heavily on the resource of the human mass, throughout Russian and Soviet military history.[127]

However, the Russian military tradition was not merely about outnumbering the enemy; first and foremost it was concerned with overcoming the enemy qualitatively—morally. According to the Russian culture of war, battles are won by men, by spiritual and psychological power, and not by machines, technology, or any other material factors. The norm codified in the eighteenth century by Marshal Suvorov glorified the endurance, patience, and bravery of the Russian soldier. In coining the adage, "the bullet is a fool, but the bayonet is a fine fellow," Suvorov was asserting that what mattered most in war was spiritual power and the troops' fighting spirit. Consequently, hand-to-hand combat with bayonets was regarded as the medium in which the powers of armies are manifested and in which those who are morally superior prevail.[128] The Russian military tradition held that Russian soldiers were "blessed" with compensatory moral qualities that allowed them to fight with inferior equipment yet to prevail nonetheless.[129] They were considered to be capable of out-suffering, outlasting, and eventually outfighting militaries of other nations. Indeed, throughout history, Russian soldiers demonstrated impressive tenacity. Brutally enforcing discipline, demanding extremes of self-sacrifice even in completely futile engagements, and insisting on subservience of the mind, body, and spirit, Russian and Soviet commanders built forces inured to mortality and suffering.[130]

With the transformation of warfare following the industrial revolution, and after every military debacle, some military specialists warned that this romantic culture of warfare—based in large part on the expendability, endurance, and courage of its conscripts—was no longer adequate to fulfill the demands of modern warfare, which posited technological mastery as a precondition for victory. However, notwithstanding the technological procurement and military modernization, the Russian military never became technocentric.[131] On the one hand, this can be explained by the traditional technological backwardness of Russia and the USSR; on the other hand, it seemed simply to reflect a military romanticism that was never fully discredited. When Soviet military thought had already become "mechanized," advanced technology never held complete sway over considerations of morale. Although convinced of the importance of technology and machines for modern warfare, Soviet military leaders believed that technology by itself was "lifeless," divorced from issues of morale backed by scientifically developed doctrine.[132]

In frames of Russian strategic culture, advanced weapons were generally regarded as "mass multipliers," and not as means to fight better with fewer numbers.[133] Even in the framework of the attempts to increase the technological quality of warfare, the component of morale was always primus inter pares,[134] and it maintains similar status in Russian military tradition to this very day.[135] The Russian reliance on limitless manpower and the emphasis on the morale factor encouraged a high tolerance for casualties and was clearly reflected in the "Russian way of war."[136] The military approach of the White and Red forces during the Civil War, Soviet fighting patterns in the Great Patriotic War, Soviet engagements in Afghanistan and Russia in the Northern Caucasus—all exemplify this tendency. This cultural tolerance of casualties was also consistent with relative indifference to the living conditions of the troops, to their equipment, and to the level of human comfort built into Soviet tanks, ships, and aircraft.[137] This traditional narrative galvanized a sense of "strategic patience" among the Russian people, and low expectations with regard to technology. The concept of time and future in Russian tradition is not measured by days or months, but by decades or generations. In contrast to Americans, a long history of wars and disasters produced among Russians a shared skepticism that a brilliantly executed campaign by advanced war machines could take the place of soldiers' long and persistent efforts.[138]

Doctrine, Technology, and Weapons Procurement

Though Marxism was strongly oriented toward materialism, fascination with technology never became a dominant part of Soviet strategic culture. There were three interrelated reasons for this reticence toward techno-euphoria. First, Russians had traditionally suffered from technological inferiority; lacking technology, there was not too much room for a fascination. Second, fascination with technology is often driven, among other factors, by casualty aversion, when technologically advanced weaponry is used as a substitute for vulnerable soldiers. This was not a consideration in the Russian military tradition, which preferred utilizing masses of spiritually superior flesh over machines. A third factor that minimized inclination toward techno-euphoria originated in the Soviet period. The regulations of weapons procurement thwarted any predisposition toward technological romanticism in military affairs and constituted the most potent factor in the Soviet attitude to technology.

Donnelly's *Red Banner* is one of the best sources for understanding why fascination with technology never became a dominant factor in the Soviet way of war. In keeping with the Russian culture of war, Soviet military science developed a distinct field that addressed the "interrelation between human being and military technology" and proved the superior role of the former over the latter. In accordance with the overriding principle that practice should be driven by theory, doctrines and concepts of operations were formulated first, and appropriate force structures were subsequently designed. Only at the end of that process was it identified what sort of technology the industry should develop and produce to satisfy the demands of the military. This deductive approach was, for the most part, fundamentally different from the American practice in this area. There, technological initiatives usually originated in industry, and the armed forces generally procured what industry sought to sell in the market. In contrast, in the USSR the driving force for technological innovations and weapons acquisition was not industrial sales or market forces, but consumer requirements. This Soviet modus operandi between the military and the industry in the field of weapons procurement was uneconomical according to free-market philosophy but reflected precisely operational necessities deduced from the nature of war.

The Soviets never denigrated the importance of technology. However, they eschewed technology for technology's sake and the notion, prevalent in the West, that weapons that were not state-of-the-art were inadequate. When opportunities emerged for creating radically new weapons utilizing technological innovations, the procurement philosophy kept technology in its proper place. To ensure that a customer's needs were satisfied, military representatives were attached to all of the defense industry design bureaus. They acted as a link between all the elements of the procurement system, from doctrine to the deployment of weapons systems. Any attempt on the part of civilian designers to introduce state-of-the-art weapons that did not fit the concept of operations as codified in the current military doctrine was ruthlessly suppressed by the military representatives. Military doctrine laid down the requirements for weapons types, and everyone involved in the process of the research, development, and procurement had a unified and standardized view of what was needed. Innovation was carried out through a directive from above.[139]

The military representatives were matching weapons to practical needs rather than pursuing technological sophistication. Under this approach to defense procurement, free from high-tech bias, no new exotic technology was produced unless it was deemed necessary by the military. These totalitarian procedures enabled the Soviets to translate the latest technological achievements to the most relevant weaponry systems in accordance with their doctrines, and protected them from a situation in which techno-euphoria usurps the development of military art. When faced with a problem, the Soviet instinct did not immediately seek a technological solution.[140] In cases of strategic competition over weapons systems, operational and conceptual creativity compensated for technological inferiority. The Soviets repeatedly preferred not to go head-to-head, but to exploit adversaries' weaknesses and to oppose them with their strengths, not necessarily in the technological realm. Dubbed in Soviet military parlance an "asymmetric response," the Soviets saw this ability to overplay Western technology through operational creativity as a form of conceptual superiority.[141] Based on dialectical laws, Russians viewed the technological superiority of one weapons system over another as ephemeral, counterweighted by the future introduction of newer technology.[142] Consequently, in accordance with principles of Soviet military science, only when integrated with all other branches of the military would advanced technology contribute to victory.[143]

Scientifically Organized Society and Soviet Military Science

Historically, Russian intellectual tradition has manifested a deep-seated propensity for self-conscious conceptualization.[144] According to Isaiah Berlin, Russian society had been never as deeply and exclusively preoccupied with constructing and learning itself as during the Soviet period. Though similar self-consciousness existed in other countries, in Soviet Russia it went further.[145] Soviet intellectuals, as "engineers of human souls," had to elucidate social goals on the basis of scientific analyses of history and then construct the society as an efficiently functioning mechanism.[146] Bolsheviks proclaimed that being should not determine consciousness, but vice versa, and set into motion the "scientific reconstruction" of all spheres of social and cognitive activity.[147] Sovietologists frequently used the term "state of the scientists" to describe Soviet civilization.[148]

The military field was not an exception. The Bolsheviks proclaimed the necessity of organizing it scientifically. With the establishment of the Soviet state, the Bolsheviks were determined to produce unique, proletarian military science. They assumed that armed with the laws of dialectical materialism, the Red Army would possess a master key to military reality.[149] Although the Soviet military theoreticians of the 1920s—Tukhachevsky, Shaposhnikov, Frunze, Svechin, and Trotsky—had different professional and political views, they all shared the conviction that "scientifically" developed military thought was a basic necessity for formulating strategy and executing tactics. Institutionalizing didacticism of military thought by introducing a unified military doctrine was not merely lip service to revolutionary ideology, but a genuine conviction that this was the only effective way to guide military practice.[150] The methodology of analysis codified in those years started with identifying the influence of technology on the nature of future war, then turned to enemy means, and finally discussed required countermeasures.[151]

Military science was an all-encompassing discipline with several subfields that regulated professional practice from strategy and doctrine, on the national level, to tactics on the battlefield. It was believed to have discovered the objective, almost natural laws of war and to have developed stratagems for military operations. Military experience was considered the essential ingredient of professionalism, but the primacy of developing theory had been established by the Soviet tradition as a norm.[152] Following the Great Patriotic War, a systematic thinking approach in military affairs was consciously constructed by the Soviets. The "culture of military thinking" consisted of numerous laws, detailed methodologies, and scientific postulates that regulated all aspects of intellectual activity in military affairs. It was the methodological polestar for how to think about the war in a scientific fashion.[153]

Forecasting Revolutions in Military Affairs

To envision the nature of the future war was the primary goal of Soviet military science. Specifically, it obliged Soviet military theoreticians to attend primarily to emerging discontinuities in military affairs—those fundamental changes taking place in operations and organizations under the impact of the new technologies.[154] According to dialectical logic, the events in military affairs did not proceed in direct causal sequence, but by means of a prolonged struggle between the conflicting trends, which finally collided at a critical stage. When in the course of the collision, thesis and antithesis eliminate each

other, history takes a "leap" to a new level where this dialectical process re-
plays itself.[155] Soviet military science treated these qualitative leaps in warfare
development as manifestations of the emergence of a new military regime. To
conceptualize the scientific-technological breakthroughs that bring radical
shifts in ways and means of waging war, the terms *Revolution in Military Affairs*
and *Military-Technological Revolution* were introduced in Soviet military
science.[156]

The primary task of Soviet military theoreticians was to determine the
indicators of these discontinuities. The prediction was carried out through a
special subfield of military science called "forecast and foresight"—the process of cognition of possible changes in the area of military affairs. "Scientific
forecast" comprised three specific tasks: envisioning the direction of military-
technological progress and the appearance of qualitatively new types of armaments; determining their impact on the emerging nature of future war; seeking methods to adjust the concept of operations, structure of the armed forces,
and weapons development to the new military regime.[157] Applying the methodology of forecasting and foreseeing, the Soviets systematically analyzed
emerging technologies to identify them as either revolutionary or evolutionary with regard to the nature of war. In fact this "scientific" prediction constituted the starting point for all military activity.[158] The default mind-set of the
Soviet theoreticians was to look for the change.

The belief in an ability to anticipate and interpret emerging revolutions in
military affairs constituted for Soviets a sense of intellectual omnipotence.[159]
In contrast to the tactical level, Soviet military science and operational theories required expecting the unexpected, being prepared for being unprepared,
and developing intellectual habits for the nonhabitual. A "sharp turn" was to be
as expected by the Soviet military theoretician as was a continuum of the current state of affairs.[160] The "culture of military thinking" prescribed refraining
from mechanical extrapolation of existing trends into the future, applying
laws of unity and the struggle of opposites, and seeking out the root causes of
change in forms and means of warfare.[161]

Authoritarian Tradition, Top-Down Organization, and the Role of the Soviet GS

Identification with authority and the wish for a strong leader have characterized the Russian people over the centuries, despite their bloody history. The
population saw authoritarian power as the antidote to helplessness and as a

THE SOVIET MILITARY-TECHNICAL REVOLUTION 49

means of guarding against chaos.[162] The inclination to be guided by strong leadership had significantly influenced organizational and managerial culture in Russia and in the USSR. Power, authority, and initiative in Russian or Soviet organizations, military and civilian alike, came from the top of the organization.[163] In Soviet society as well, everything was controlled, integrated, and organized hierarchically. The authoritarian center was the primary agent initiating social, economic, and technological change. Military writings, plans, innovations, and international negotiations nearly always began with a "top-down" argument.[164] This authoritarian cultural legacy was fully manifested in the role the General Staff played in Russian and Soviet military traditions.

The Soviet GS was the most central, important, and influential military institution. Soviet military theoreticians envisioned the GS as an elite and intellectual avant-garde of the defense establishment and dubbed it the "brain of the military." The main task of the GS was not only to plan and execute operations, but first and foremost to synthesize insights of military science and develop fundamental military knowledge.[165] Borrowing the concept and the organization of the GS from Prussian military tradition, in the Russian and later the Soviet armies, a chosen few from the officers' corps formed the GS. They saw themselves and were seen by others as the crème de la crème of the military elite. In line with Prussian heritage, being a GS officer was a military profession in itself. The most prominent and capable officers were carefully selected from the entire corps and put through the most prestigious training program in the military.[166]

Soviet military theoretician Gen. Aleksandr Svechin argued that the main goal of the GS was to generate *revolutions in military affairs* and to determine how they can be implemented. Similar to the Russian scientists, the military theoreticians insisted that the GS should derive its intellectual strength not from specific knowledge in a certain field but from its all-encompassing, encyclopedic grasp of any given military issue. Svechin demanded that officers of the GS be "encyclopedists." He expected them to be able to distill knowledge by conceptualizing from the general to the particular.[167] The curriculum and military career that every officer went through prior to his arrival at the "brain of the military" prevented him from developing a parochial mentality with loyalty to a specific service. The Soviet GS officers were driven by synthetic, nonpartisan, and genuinely joint considerations when they developed military science, thought through the future of warfare, made procurement decisions, or planned military operations.[168] The support pillars for the "brain of the military"—four of the ten departments of the Soviet GS—were directly and continuously involved in developing knowledge concerning the nature of future war.[169]

Notwithstanding the glorification of the combat heritage in the Soviet military tradition, a biography that included combat excellence was neither necessary nor sufficient to become the chief of the GS, as a battlefield experience was not a prerequisite for becoming a GS officer. Marshal Ogarkov served most of his career as a military engineer in various command positions. Though a participant in WWII, he never commanded a fighting division. His nomination as chief of the GS, however, did not arouse any professional opposition. According to Generals Gareev and Varennikov, two deputies of the chief of the GS, and both officers with considerable combat credentials, Ogarkov was unparalleled in bringing the GS to its functional ideal—turning it into the real brain of the military.[170]

The chief of the GS was considered to be the leading defense intellectual of the Soviet Union and played a pivotal role in the doctrinal development of the military. It was not enough for him just to manage all the combat forces; he was expected to be the primary deductive thinker in military affairs. Throughout Soviet history several chiefs of the GS were well-known military theoreticians: Mikhail Frunze, Mikhail Tukhachevsky, Georgii Zhukov, Vasilii Sokolovsky, and Nikolai Ogarkov.[171] It is emblematic of this tradition that at the moment in the 1970s when the Soviet GS realized that the Revolution in Military Affairs was under way, it ordered publication of a new military encyclopedia. Its eight volumes were written between 1976 and 1980.[172] In keeping with established norms, Chief Ogarkov contributed a number of entries, with the best-known articles related to the theory of the Military-Technical Revolution.[173] That being said, the GS was not a futuristic crowd of military theoreticians, but a "brain" that commanded operations that it had planned singlehandedly and coordinated ground, air, and naval forces on a number of fronts. It envisioned future war and then prepared, planned, and executed it.[174]

Time Orientation and Words versus Deeds
Such interrelated features of Russian-Soviet culture as the orientation toward an imagined future, a profound dissonance between words and deeds, and a pathologically inept implementation of ideas were strongly manifested in Soviet military affairs. Rarely were the holistic and profound visions of Soviet theoreticians transformed into concrete, practical deeds.[175] A detailed analysis

of these traits lies beyond the scope of this research, but some observations in this regard will be presented here.

Russians think in terms of expanses of time and space that differ from those of most of their Western counterparts.[176] In contrast to a practical US predisposition to be anchored solidly in the present, Russian cultural orientation is toward the past and the distant future.[177] Anthropologists argue that Russian society possesses more irrationalism than pragmatism; and, applying Jung's theory of psychological archetypes, scholars classify Russian society as being closer to the intuitive-sensitive than to the verbal-logical type usually attributed to Westerners.[178] The traditional Russian tendency to exist in the imagined future was reinforced by Marxist dialectics.[179] The attempts by *homo Sovieticus* to construct a desirable future often metamorphosed into a grotesque falsification of reality and faith in it.[180] This cultural tendency to exist in the imagined world was fueled by the Russian tradition of staging an event for show for officials and also for self-consumption. Known historically as the phenomenon of "Potemkin villages," it is a tradition that materialized in its Soviet variant as *pokazukha*.[181] "Compensatory symbolism"—a propensity to exaggerate achievements or to fabricate reality—is a Russian cultural-psychological trait both in civil and in military affairs.[182]

Discussing Soviet strategic culture, Fritz Ermarth points out Russians' emphasis on their "spirituality."[183] The messianic idea of harmonious spiritual community—*sobornost'*—when Russian people see Moscow as the Third Rome, was cultivated by Russian rulers for about a thousand years.[184] Influential in internal and foreign policy, this narrative is similar to the American notion of being a "city upon a hill." However, the by-product of Russian ethnocentrism was very different. Seeing the soul and not the mind as the central component of being,[185] Russian tradition cultivated a self-image of being less inclined toward active pragmatism. It was seen as inconsistent with a normative emphasis on the spiritual versus the material aspects of life.[186] Coupled with the continuous negative historical experience that favored low expectations about what can be achieved, it thwarted a tendency toward optimistic practicality or rational action in Russian tradition.[187] While typical American popular heroes usually act in order to find truth and to impose justice, leitmotifs of Russian literature rarely go beyond posing the age-old rhetorical questions of "who is to blame?" and "what is to be done?"[188] Problems are dealt with by ignoring them.[189] As one psychiatrist argued, Americans are oriented toward doing, Russians toward contemplating.[190] A Soviet physicist, P. L. Kapitsa, argued

that fixating on generating holistic, fundamental theories made Russians traditionally good in theory, but extremely bad in implementing ideas.[191] The
Russian-Soviet focus was usually on the wishful outcome, rather than on the
starting point.[192]

The gap between the theoretical and the feasible never stopped Russians.
In the early 1970s cybernetics as a field of study enjoyed far more prestige in
the Soviet Union than elsewhere. Computer production, however, both qualitatively and quantitatively lagged far behind the United States. With obviously underdeveloped hardware and software, the Soviet thinkers generated
very advanced theoretical insights on technology applications and the role
computers would play in the future.[193] However, they never implemented
them. The same cultural features were manifested in frames of mind and managerial activities in Soviet military affairs. Russian and Soviet military thinking style was future-oriented and repeatedly manifested wishful thinking based
upon expectations while ignoring current realities and neglecting problems.
The Soviets were traditionally good at theorizing innovative concepts, but pathologically bad at implementing them. They often limited themselves to abstract
considerations and remained the prisoners of their futuristic visions—not connecting words and deeds. Theories and even training were about future war—
the war that could take place after weapon plans had been fulfilled. Announc-
ing a project gave it a reality—even if it would be years before it was actually
realized. Military theoreticians often debated futuristic visions without concrete and complete calculation of their feasibility.[194]

Soviet military thought often became a function of abstract ideas, where
sophisticated doctrines were incompatible with the country's operational capacity to implement them. It often responded to what would become theoretically possible when new weapons became operative. Admiral Makarov went
into the Russo-Japanese War to command an obsolescent fleet, which was trying to execute advanced and creative tactics. Soviet "deep battle" theories
adopted by the Red Army in the 1930s went far beyond Soviet technological and
economic capabilities of that time. Khrushchev made missile threats according to plans that the Soviet space research and industry had only on its draw-
ing boards.[195] This cultural tendency had its pros and cons. On the one hand,
despite the impressive amount of thought given to military science, many of
the ideas discussed were infeasible.[196] On the other hand, this thinking was
not restricted by the frames of objective reality. Consequently, Russians often
thought "outside the box" about the unthinkable, with innovative and creative

visions of ways to achieve victory. Though at times unrealistic, their intellectual products were competitive, and their creativity often helped them to overcome their counterparts in other, wealthier countries.[197]

In referring to the failure to implement the MTR, an additional cultural obstacle must be mentioned as well. The operational concepts of the MTR were at odds with the norms of the Soviet military. The tactical creativity, flexibility, and capacity for independent action so vital to operating MTR-era weaponry were poorly developed in the Soviet military, which was educated in adherence to book solutions, detailed planning of operations, centralization, and limited decision making permitted to junior officers.[198] Although a great deal of progress was made in those years in improving the mental flexibility of Soviet junior commanders, initiative on the tactical level was a very difficult concept for the Soviet military to come to terms with.[199] At the same time, this independent and improvising state of mind and delegation of authority to tactical levels were essential conditions for successful implementation of the MTR concepts. The rigidity of the Soviet command and control system made it harder to implement this RMA, which implies the establishment of decentralized small units that have their own independent means of collection and destruction.

To make things worse, the Soviet military profession never considered tactics prestigious and noble, and never emphasized them, relative to other levels of war. Soviet military publications paid marginal attention to tactical versus operational or strategic issues. *Voennaia Mysl'* periodically analyzed Western tactical innovations but paid no attention whatsoever to discussing modern Soviet tactical practices. "Tactics" were considered only marginally important, and a separate section of *VM* was devoted to them beginning only in the 1980s. The tactical magazine *Voennyi vestnik* focused on techno-tactical issues, with no theoretical or fundamental discussions.[200] The *Soviet Military Encyclopedia* as well as the classic *History of the Great Patriotic War* barely discussed the tactical issues of military art. In an atmosphere that did not allow open criticism of senior commanders, the successful tactical experience of WWII was glorified, studied, and its lessons applied dogmatically. Tactical failures, however, were obscured and ignored. All this was reflected in the conduct of the junior Soviet officers. Since the late 1980s, prominent Soviet military theoreticians published introspective articles that criticized all the Soviet tactics for conservatism, dogmatism, intellectual stagnation, the lack of junior officers' creativity, innovativeness, and ability to improvise.[201]

CONCLUSION: CULTURAL FACTORS AND
THE SOVIET MTR

Soviet MTR concepts represented a form of theoretical conceptualization that chronologically preceded technological procurement, combat experience, and the capability to implement it. The Russians coined the term *Revolution in Military Affairs* and produced pioneering theoretical works without possessing either weapons or technologies. Owing to their technological inferiority, the Soviets relied on Western scientific advantage and exploited it as a conceptual frame of reference for their own doctrinal innovations. This study maintains that the bulk of the explanation for why the technologically inferior USSR prevailed qualitatively over the West in the conceptualization of the emerging military-technological realities lies with several qualities of Soviet strategic culture that contributed to more rapid comprehension of the emerging discontinuity in the nature of war.

The traditional Russian cultural predisposition to holistic inferential procedures fostered a synthetic thinking by Soviet military theoreticians. For centuries, Russian scientific and intellectual circles were dominated by the belief that all observed phenomena were organically interconnected and that every element was in constant interplay with others in frames of an indispensable and synthetic metasystem. Holistic-dialectical cognitive style—a predominant mental formation in frames of the Soviet strategic culture—inclined to relate to the context as a whole, and to assign causality to the relationships between a focal object and the field. The achievements of scientific-technological progress stood for the *focal point,* while the changes in the ways and means of military operations stood for the *field.* The focus on dialectical change made it more natural for the Soviets to think in terms of discontinuities and to pay attention to the RMAs. The latter constituted the potential transformation in the relationship between technology on the one hand, and organizational and operational concepts on the other. Thus, the introduction of weaponry, communications, and intelligence sensors associated with microprocessors swiftly became organically connected in the Soviet mind with its impact on the nature of war. This way of thinking made it easier to conclude whether the new technology constituted a continuation of the existing trend in military art or had brought a revolutionary transformation in military affairs.

Soviet military science institutionalized this holistic-dialectical approach to military affairs. An intellectual atmosphere established in Soviet professional circles fostered the diagnostics of discontinuities in the nature of war. Soviet military science studied the ways to construct and the means to operate military machines in a modern war, discovered the objective, almost natural laws of warfare, and deduced stratagems for military operations. The Soviet military developed thorough methodological guidelines on how to think about war scientifically. Military theoreticians were obliged to envision the nature of future war, and to determine the indicators of the emerging discontinuities in military affairs. Applying forecasting methodology, the Soviets systematically analyzed the emerging technologies to identify them as either revolutionary or evolutionary and to assess their impact on force structure and concepts of operations. By the early 1970s, Soviet military theoreticians had identified two periods of fundamental change in military regimes during the twentieth century. Thus, when the advanced sensors, command and control capabilities, and PGMs started to emerge, Marshal Ogarkov and hundreds of Soviet military forecasters already possessed the advanced system of terminology and cultural aptitude to comprehend the paradigmatic change in military affairs. Given their technological inferiority, the Soviets successfully exploited the Western scientific advantage as a conceptual starting point for their diagnostics and eventually identified a new Military-Technical Revolution.

The nature of the Soviet defense and military bureaucracy made it possible to channel the findings of military science into concrete organizational or conceptual transformations. In keeping with the authoritarian cultural legacy and centralist Russian mind-set in management, Soviet institutions were controlled, integrated, and organized top-down. The GS, accordingly, executed the highest authority in all existing aspects of Soviet military affairs. It played a pivotal role in developing a new theory of victory and created the institutional framework for generating knowledge on the Revolution in Military Affairs. The Russian/Soviet tradition envisioned it as the "brain of the military." Consistent with this definition, its role was not only to plan and execute operations, but first and foremost to synthesize insights of military science and develop fundamental military knowledge. The role given to the GS in the Soviet military system, the professional approach of the GS officers, and the intellectual atmosphere in its directorates enabled this brain of the military to capitalize on its intellectual potential and to develop constantly fresh military ideas about the future of war and then to distill from them strategy, doctrine, and weapons development. The GS possessed a unique triumvirate of absolute authority: to forecast future trends in the theory and practice of war, to experiment with new operational and organizational concepts, and to initiate the innovation from the top of the organizational pyramid through all the branches and services of the Soviet military. It played a far more central role in initiating military innovations than did similar organizational institutions in the West.

When the new weaponry began to emerge during the 1970s, a long-established disinclination of the Russian military tradition from technological romanticism kept Soviet military experts from concentrating on the narrow aspects of the newly introduced weaponry but favored the broader conceptualization of the relationship between technology, organization, and doctrines. Being for centuries materially inferior, and possessing a low capacity to equip the military with modern weaponry, left little room for techno-euphoria. Moreover, the traditional cultural narrative, which emphasized endurance, self-restraint, and moral and physical fortitude, was infused in Russian military culture. Consequently, the strategic belief of the "Russian way of war" was that battles are won by men, spiritual power, and psychological factors, and not by machines, technology, or any other material component. This romantic "spiritualism" was never discredited. While Soviet military thought was already "mechanized," advanced technology never held total sway over morale factors and indigenous operational concepts. Casualty aversion, another potential driver of techno-euphoria, never was a consideration in Russian military culture. The Soviets regarded technology as "lifeless" if divorced from human, operational, organizational, and doctrinal factors. In keeping with the norms of Soviet military science, technology development and weapons procurement in the USSR were driven by military doctrine. The primacy of military art over technology was established as a professional norm. The military invested in weapons to perform the role it was given in military doctrine. This philosophy kept technology in its proper place and prevented techno-euphoria from usurping the development of military art. Together with other factors, this centralized, authoritarian approach prevented techno-romanticism and facilitated the translation of technology into the most relevant combat architecture in accordance with established military doctrine.

The first to pay attention to the discontinuity in military affairs and a pioneer in developing revolutionary concepts of modern operations, the Soviets never implemented their theories. This state of affairs was repeated so often in Russian military history that it can justifiably be seen as a feature of Russian strategic culture. Russia is a being culture, not a doing culture, and it has a strong inclination toward declarative knowledge. Understanding an issue is more important than action. The bearers of Russian-Soviet strategic culture were traditionally good at theorizing about innovative concepts but pathologically bad at executing them. Russian military thought and Soviet military doctrines often became a function of abstract ideas, where sophisticated concepts of operations were incompatible with the country's capacity to accomplish them. The implementation of MTR concepts and massive fielding of MTR weaponry was beyond the political, economic, and cultural capacity of the Soviet state. However, since the imagination of the Soviets was not restricted by the frames of objective reality, they frequently thought "outside the box" about the unthinkable, and emerged with innovative and creative theories of victory. The MTR story serves as a good example. Though unrealistic for them to implement, its intellectual products were competitive and often surpassed their counterparts in creativity.

3 THE IMPACT OF CULTURAL FACTORS ON THE US REVOLUTION IN MILITARY AFFAIRS

ALTHOUGH the American technological capability to execute deep strikes using PGMs was significantly more advanced than that of its Soviet counterpart, the Soviet military had fuller comprehension of the revolutionary impact that the ALB and FOFA arsenals would have on the future battlefield than did the US military.[1] It took the Pentagon almost a decade to "become converted to the MTR" and to emulate the Soviet proposition of major changes in the character of warfare, which saw the "reconnaissance-strike complex" as the dominant architecture for future operations.[2] The phase of American "capabilities formation" was followed, beginning in the late 1980s, by the conceptual birth of the RMA. Not until Andrew Marshall and his colleagues introduced the notion of the RMA did this conceptual innovation reach the consciousness of the American military and defense establishment. The purpose of this chapter is to trace the impact of the American strategic culture on the approach of the US defense community to the RMA.

The chapter comprises three parts. The intellectual history of the American RMA is discussed in the first part. It describes the technological and conceptual preconditions for the innovation, touches on the US intelligence assessments of the Soviet MTR, refers to the conceptual birth of the American RMA, and analyzes the implication phase of this innovation. The second part concentrates on the general sources and the main traits of the American strategic culture. It describes several fundamental societal and cognitive characteristics and illuminates the principal traits that gave shape to American strategic culture, such as: the "American way of war"; "a-strategic thinking"; the engineering approach to security; time orientation; the role played by the Joint

Chiefs of Staff; a role of technology in military affairs; and inclination to eth-nocentrism. The conclusion integrates the previous two parts and uses the characteristics of the American strategic culture to account for the conduct of the US defense community with regard to the RMA. The analysis of the cultural traits clarifies why the cultivation of the technological seeds of the American RMA preceded the maturation of the conceptual ones, and what prevented the US defense community from grasping the emerging RMA in a timely and proper manner.

PART ONE: AN AMERICAN REVOLUTION
IN MILITARY AFFAIRS

The Formative Period of the RMA: Technological
and Conceptual Preconditions (mid-1970s–early 1980s)

According to William Owens, one should associate a technological prequel of the American RMA with the Pentagon officials who began in the late 1970s to think about the application of technology in military affairs and to formulate a so-called offset strategy[3]—a program by which the US and NATO allies could use technological superiority to neutralize the overwhelming conventional-forces size advantage that the Warsaw Pact members had over NATO forces in Europe.[4] William Perry, undersecretary of defense for research and engineering, responsible for the development of the capabilities for the "offset option," stated in 1978: "Precision-guided weapons, I believe, have the potential for revolutionizing warfare. More importantly, if we effectively exploit the lead we have in this field, we can greatly enhance our ability to deter war without having to compete tank for tank, missile for missile, with the Soviets. *We will effectively shift the competition to a technological area where we have a fundamental long term advantage* [emphasis mine]." Although in retrospect Perry claimed that the offset strategy was more than just a plan to exploit "high technology" for its own sake, the primary objective of the defense establishment was to use high technology to build better weaponry systems than those of the Soviet Union.[5]

The means of precision strike, intelligence, and communication—the capabilities on which the concepts of the American RMA would later be built—matured technologically in various Defense Advanced Research Project Agency (DARPA) projects beginning in the late 1970s. DARPA allocated its budget to give qualitative advantages to American forces to offset the quantitative superiority the Soviet forces enjoyed in Europe[6]—the strategy pursued during the

1970s and 1980s.[7] Among the core technological focuses that shaped re-search, development, and production efforts were: *the families of land-, air-, and sea-launched precision-guided and stand-off weapons* such as terminally guided submunitions, smart and guided missiles and bombs, and stand-off land attack missiles; *command and control and automated reconnaissance and target acquisition projects* such as Airborne Warning and Control System (AWACS), E-8 Joint Surveillance and Target Attack Radar System aircraft, Integrated Targets Acquisition and Strike System, and stand-off target acquisi-tion system (SOTAS); programs to bolster *anti-armor weapons,* such as "fire and follow" and "fire and forget" antitank missiles; *navigation and guidance devices* based on the global positioning system (GPS); *stealth technology* such as F-117 aircraft or naval stealth and stand-off precision strike capabili-ties; and *unmanned aerial vehicles* of various altitude and endurance.[8] In 1978 DARPA integrated research and development of the above-mentioned capabilities under one unified project entitled "Assault Breaker." From the 1980s, the project became also known as the "Smart Weapons Program." Tomes defines this period of capabilities-developing as the "formation" stage of the RMA.[9]

Air-Land Battle doctrine, which was invented to stop the Soviet second ech-elons deep in the rear, laid down the technological fundament of the future American RMA.[10] Its more important contribution, however, was probably the unprecedented introduction of the operational perspective to American mili-tary thought, or what Shimon Naveh defines as the emergence of "operational cognition."[11] In his landmark works on military history, Erickson claims that the recognition of the operational level is necessary to "think big" about war. To him, operational art is a means of accommodating technical change to pro-duce new warfare concepts.[12] Consequently, the introduction of operational perspective became the conceptual precondition for the American RMA.

There tends to be agreement among scholars that American receptivity to operational thinking in the late 1970s was promoted by the poor perfor-mance of the US Army in Vietnam.[13] Not before the early 1980s, as Lock-Pullan shows in his study, did US armed forces start to make the conceptual leap into the operation level of thinking, when they embarked on the ALB doctrine.[14] During this "reawakening of American military thought, the operational level of war became a key focus of study and an important consideration in defense planning."[15] In addition to other sources of inspiration,[16] Soviet op-erational theories stimulated more sophisticated and systematic thinking by

American theoreticians about the nature of battlefield integration and extension. ALB mirrored many of the developments of Soviet operational theory
from the 1920s.[17] The emulation was so apparent that according to Erickson,
"Generals Svechin and Isserson and Marshall Tukhachevskii, would at once
be impressed and flattered, sufficiently so even to overlook the protracted in-
trusion upon their copyright."[18] The operational corpus of knowledge facili-
tates an intellectual climate suitable for systematic thinking about changes in
military affairs. Consequently, it is indispensable for the diagnostics of future
RMA. Without the broad professional aptitude for operational thinking in
the US military, RMA concepts would have remained an abstract idea. Tomes
concludes that the evolution of American military thought in frames of the
ALB, and specifically the introduction of the operational level of thinking,
was central to the future evolution of the American RMA.[19]

Azar Gat makes a sound argument that ALB and FOFA were devised as
early as the 1970s to incorporate the new technologies outlined above. Thus,
argues Gat, although the Soviet MTR vision was more comprehensive in understanding the revolutionary nature of the change, the US and NATO doctrines chronologically coincided with the Soviet theorizing.[20] Gat's argument
is persuasive. However, it is important to state that these capabilities and visions have not matured either technologically[21] or conceptually during the
period under discussion. For the most part, the defense community treated the
emerging capabilities as a multiplier of the existing force effectiveness and did
not deduce from it any revolutionary implications with regard to the concept
of operations, organizational structures, or the nature of war.[22] Without a deeper
understanding of the operational and organizational consequences of the new
weaponry, the mere existence of the smart weapons and technologies did not
produce the Revolution in Military Affairs. New weapons systems were produced in compliance with very mechanical logic—to ensure that the United
States was not left behind in the area of new technology.[23] It was not the futuristic vision of military thought that was the driving force behind the innovations, but the linear arms race logic vis-à-vis the Soviet adversary. The offset
strategy certainly reflected an American way of looking at the world and coping with its problems through its typical way of war. Seeking technological
answers to the operational questions of the central front reflected a cultural
affinity to leverage the challenges to national interests by technology.[24]

The key capabilities developed, then, created the technological quality that
in tandem with the sophistication of American military thought produced

the fertile soil for launching and realization of a bold defense transformation in the 1990s.[25] However, the cultivation of the technological seeds of the American RMA preceded the maturation of the conceptual ones. The offset response consisted of little more than sustaining a technological edge in the face of an armored assault by the Warsaw Pact forces.[26] The corpus of operational knowledge was a solid basis for the development of new ideas but still inadequate for generating revolutionary visions of future war. No advances in reconceptualizing the existing paradigm about warfare were made in those years. The PGMs were seen as just another weapon in the military arsenal. The very community that developed the weaponry failed to recognize their potential in future war.[27]

American Intelligence and the Assessment of the Soviet MTR

Although there were a few academicians who saw the future in the early 1970s,[28] Knox and Murray claim that the tactical emphasis of the Pentagon's analysts had prevented them from seeing anything revolutionary in these new capabilities.[29] To make things worse, the phase of the technological and conceptual preconditions of the American RMA coincided with the misinterpretation by US intelligence of Soviet MTR concepts. The disregard of the American defense community for the emerging change in the military regimes lasted for about a decade. This inattention is particularly striking in light of the wealth of information accumulated in US intelligence about the Soviet theoretical writings on the MTR.

In its analysis of the Soviet perception of Western military capabilities, US intelligence detected at a very early stage, and with a high level of accuracy, the new direction of military thought that was evolving in Soviet military circles. By the mid-1970s, the US had developed a general understanding of the mechanisms of how the Soviets developed their military concepts, including exercises to test theoretical propositions, doctrinal discussions, and scientific conferences.[30] The intelligence community translated and disseminated Soviet writings on military thought, doctrine, strategy, and operational concepts to make important information easily available across the services.[31] The CIA had at its disposal a considerable amount of open Soviet sources that reflected the intellectual debate about the emerging MTR and its implications for the Soviet vision of future war.[32] These sources, which included translations from the classified journal *Voennaia Mysl',* shed a great deal of light on the term *Military-Technical Revolution* within the context of Soviet military thinking at the time.[33] In 1974, the seminal work *Scientific-Technical Progress and the Revolution in Military Affairs* was translated and disseminated by the CIA.[34] In 1981, a special report was dedicated to the Soviet methodology of "forecasting in military affairs," which inquired into the nature of the paradigmatic changes in the nature of war and into the essence of the current MTR in particular.[35]

Since the late 1970s, US intelligence closely monitored the growing Soviet interest in microelectronics, computers, and signal processing, and Moscow's continuous efforts to acquire them by both legal and clandestine means.[36] The CIA reported conspicuous Soviet concern with regard to the technological lag vis-à-vis NATO, particularly in key technologies that provided precision weaponry capabilities.[37] The analysts argued that the Soviet search for technologies was a necessary starting point in the implementation phase of the MTR decreed by the Soviet chief of staff. They estimated that the Russians intended to use the MTR concepts, and especially PGM, to change the total force structure and combat potential of Soviet forces.[38] The "smart" precision-guided munitions, which the Soviet military reckoned would alter the nature of war, relied on a variety of technologies in the field of microprocessors and computers, and consequently their acquisition became a more urgent necessity.[39]

On the heels of monitoring the Soviet quest for advanced technology, US intelligence soon began to arrive at the operational essence of the MTR—Soviet experimentation with reconnaissance-strike and fire complexes. Discussing Soviet conventional doctrine, the CIA understood that the Soviets considered conventional weapons so accurate, lethal, and destructive as to approach the potential of nuclear munitions.[40] A series of CIA estimates from the early 1980s refer to so-called *reconnaissance-strike organizations* (RSO), which had been developed out of the Soviet concern for the threat posed by the Assault Breaker, precision-guided, deep-striking, theater-level systems capable of firing on follow-on moving Soviet echelons.[41] According to the same estimates, the RSOs were a further expression of the new MTR concept of integrated, deep, simultaneous fire destruction of the enemy. The analysts grasped that the Soviet RSOs consisted of an integrated triad of reconnaissance and target acquisition complexes, automated command and control elements, and long-range striking systems. They correctly attributed the ROK and RUK to the operational (army) and the tactical (division) levels and envisioned them as the main trend in future Soviet force development.[42]

In the late 1980s the CIA reported that since the 1970s, motivated by the need to counter NATO deep-attack, high-technology conventional weapons and extended battlefield concepts, the Soviets had been able to match NATO capabilities in nearly every major ground forces weapons category. Discussing the Soviet conventional doctrine, the CIA acknowledged Soviet declarations regarding their perception of the virtual parity of conventional versus nuclear weapons. The CIA report argued that military advantages afforded to the USSR by its numerical supremacy might be mitigated by Western progress in advanced technology conventional weapons, especially long-range PGMs.[43] Toward the end of the Cold War, the CIA attained additional clarification of the Soviet doctrinal vision, which reckoned that the outcome of future war would be determined mainly by a massed strike of conventional PGMs linked to real-time reconnaissance systems and complementary ground maneuvers, rather than by masses of tanks, infantry, and artillery.[44]

However, in forecasting the development of Soviet military power for the 1980s, US intelligence concluded with an assessment that minimized the over-all implications of the Soviet innovation. US intelligence predicted that if current trends continued, "new technology, whether developed or illegally ac-quired, was expected to lead to evolutionary improvements in individual systems. However, not one of these technological developments or even their combination in the foreseeable future was expected to revolutionize modern warfare."[45] Similarly, while discussing Soviet writings on the MTR and RUK concept during the early 1980s, senior DoD officials treated the issue according to arms-race Cold War logic: if the notion of what the Soviets termed Western "reconnaissance-strike capabilities" caused a certain strategic discomfort in Moscow, then the US should expand its investment in this area.[46] This logic was consistent with various administrations' efforts—among them economic ones—to neutralize Soviet influence, to place them at a competitive disadvantage, and to bring the struggle to an end on American terms.[47]

The wealth of information concerning Soviet views on the discontinuity in military affairs, accompanied by the poverty of comprehension regarding its consequences, was a situation that endured within most of the US defense community for almost a decade. Only a few American analysts, most notably Gen. William Odom, focused on the validity of the MTR and recognized it as more than just another Soviet innovation.[48] Most Soviet-watchers in the West, in their analysis of Soviet theoretical writings, were unable to see the forest for the trees of specific technologies and tactical-operational problems.[49]

The Conceptual Birth of the American RMA

Without a specific date for the birth of the current American RMA, one can designate the period of the late 1980s to early 1990s as the intellectual cradle of the paradigmatic change of American security thought. MacGregor Knox and Williamson Murray contend in their work on the dynamics of military revolutions that Andrew W. Marshall and his experts within the Office of Net Assessment (ONA) were the first to register the significance of Soviet writings on the MTR and to introduce the notion of revolutions in military affairs into the American defense community.[50] This claim was echoed by Gen. Makhmut Gareev, the president of the Russian Academy of Military Sciences. Gareev identified Marshall as a theoretical luminary who fully grasped the essence of the Soviet MTR and as a founding father of the American RMA.[51] Although the technological groundwork for the innovation had been laid down in the 1970s, for the American defense community the RMA thesis had been nothing but a vague, abstract term when Andrew Marshall and Andrew Krepinevich first circulated their memorandum on the RMA in the early 1990s. The US armed forces (similar to the British when they first began experimenting with armored and mechanized warfare in the mid-1920s) were not consciously thinking in terms of a revolution.[52] As one scholar remarked, the US military, like Molière's character in *The Bourgeois Gentleman,* had been "speaking in prose" (the RMA) but did not know it.[53]

Indeed, only a small group on the margins of American defense planning in the early 1980s would recognize the approaching RMA.[54] Albert Wohlstetter is generally considered to be the first senior figure within the American defense establishment to understand the dramatic impact of the new accurate weapons on the nature of war. Wohlstetter referred to the phenomenon as "revolution in the accuracies of unmanned weapon systems."[55] Envisioning the first-generation PGMs deployed in the latter stages of the Vietnam War, he realized their potential for substantial reduction of the inefficiencies and uncertainties that had plagued large-scale industrial age combat. In the face of what he called the "enormous inertia" of the armed services, Wohlstetter, supported by Andrew Marshall and a few other defense intellectuals, campaigned vigorously through the 1980s to consider more carefully the strategic implications of an expanding family of PGMs. In his view, the "revolution in microelectronics" opened up new vistas for the application of force and an increasingly wider variety of political and operational realities.[56]

It was only at the very end of the Cold War that a genuine interest in Soviet MTR theories gathered momentum in the American defense establishment. The highest point of Wohlstetter's national-level efforts to incline the defense community to reconceptualize the nature of warfare came in 1987, when he cochaired with Fred Ikle the Commission on Integrated Long-Term Strategy. By this time, it was no longer the standard intelligence analyses on the doctrinal action-reaction dynamic in the European theater that attracted American attention, but the essence of the discussion about the emerging nature of the future security environment. The report discussed the necessity of extending its studies beyond Cold War military balance assessments, even though the USSR was still alive and kicking.[57] The commission's report credited American technological advances discussed above, such as stand-off PGMs, space, "stealth," radar, and targeting capabilities. However, the report stated without hesitation that while the Soviets fully appreciated the implications of these systems on the ways of waging modern warfare, the Pentagon did not. On a more positive note, the commission declared that if the US awoke to the opportunity at hand, it might acquire a more versatile, discriminating, and controlled capability to employ this technology-driven change in war.[58]

To further develop its initial insights, in 1988 the commission established a working group, cochaired by Andrew Marshall and Charles Wolf. The group, which included a few select defense intellectuals from the establishment and academia, was entrusted with the task of projecting the likely contours of military competition in the future security environment. The report echoed the findings of its predecessor when it stated that the Soviets had identified roughly the same list of technologies important for future war but had considered their implications more systematically. It stated further that most, if not all, considerations given to this subject in the West had focused too narrowly on the utility of highly accurate, long-range systems for raising the nuclear threshold and enhancing conventional deterrence.[59] According to the Marshall and Wolf report, rather than merely identifying ways to improve specific systems or perform existing missions, Soviet writings had suggested that the conduct of war would be broadly transformed by a "qualitative leap" in military technologies. The report found that in contrast to the American approach, the Soviet MTR writings tended to focus not on questions of feasibility, cost, and timing for specific innovations, but rather to assume that families of new technologies would eventually be introduced, and to examine the tactical, operational, and strategic implications of technological trends. The report asserted that the Soviets envisioned a more distant future than American military experts, and allowed that the Russians might be correct in their assessment that the advent of new technologies would revolutionize war. The group concluded that if this was indeed the case, then a transformation in the nature of war would affect American force structures and command practices in some cases more profoundly than the introduction of nuclear weapons.[60]

From the late 1980s, Andrew Marshall eclipsed Wohlstetter as the leading proponent of inquiring into a potentially emerging paradigmatic change in the future security environment. Building upon its work for the above commission, ONA embarked on a more detailed assessment of the Soviet MTR vision, starting in 1989. The preliminary lessons from the Gulf War provided further stimulus for this inquiry, as the US sought to conceptualize the new type of warfare seen during Desert Storm. The US specialists claim, and the Soviets concur, that during Operation Desert Storm the allies successfully executed a perfect version of the Soviet conventional theater offensive, which encapsulated most of the doctrinal principles developed by Soviet military theoreticians in frames of the MTR. In Ogarkov's view, the most impressive allied capability demonstrated during the war was the ability to conduct a tightly synchronized, integrated joint operations assault throughout the depth of the operational theater, striking both the enemy's strategic centers of gravity and its military forces, in order to produce decisive results.[61] The ONA experts had picked up on the writings by the Soviet military and offered an assessment that had two related goals: first, to identify whether or not the Soviet analysts were correct in their conviction that they were witnessing a fundamental discontinuity in military affairs; and second, if a military revolution was indeed on the horizon, to pinpoint critical issues that had to be given a prominent place on the defense management's agenda.[62]

This assessment of the Soviet MTR, which was completed in 1992 (with a more comprehensive assessment a year later), is perhaps the best-known document prepared by ONA. The ONA intellectual effort yielded what seemed to be a total consensus that Soviet theorists had been correct since the late 1970s about the character of the emerging MTR. The net assessment confirmed the Soviet postulates that assumed that advanced technologies, especially those related to informatics and precision-guided weaponry employed at extended ranges, were bringing military art to the point of revolution in the nature of warfare. Along with *information warfare,* the report identified the Soviet concept of *reconnaissance-strike complexes* as the main determinant of future warfare.[63] The 1992 and 1993 assessments called for a significant transformation of the American military, not so much in terms of new technologies but rather in operational concepts and organizational innovation. Being more advanced in these two fields was expected to be far more enduring than any advantage in technology or weapons systems. The report underscored the importance of a concept of operations in identifying the most effective weapons. It criticized a tendency to utilize the advanced technologies simply as "force multipliers" in frames of the existing approach to warfare. The assessments attributed the highest importance to the investigation of and experimentation with novel concepts of operations and deducing from them a new architecture of military power.[64]

In contrast to the traditional "technology-driven" mentality of the American defense community, Andrew Marshall and his experts emphasized above all the conceptual and doctrinal, rather than the purely technological, aspects of the RMAs. The memorandum stated outright that although one would clearly want to have superior technology, the most important competition is not the technological but the intellectual one. The main task is to find the most innovative concept of operations and organizations, and to fully exploit the existing and the emerging technologies.[65] The phrase MTR denoted too great an emphasis on technology, and therefore an alternative term, *Revolution in Military Affairs*, was adopted. It is interesting to note that this expression as well was borrowed from Soviet military writings on the subject, though ONA experts considered it preferable because it emphasized *revolution* rather than *technology*.[66] According to William Owens, then vice chairman of the Joint Chiefs of Staff (JCS), Soviet ideas regarding the MTR had stirred enough interest among observers of Russia in the West to reduce it to the official Pentagon acronym. "A higher form of praise of Pentagon officials does not exist."[67] The observations about the characteristics of a new Military-Technical Revolution were made on the basis of Soviet and Russian insights presented in their writings and personal exchanges with Soviet/Russian specialists during the early 1990s.[68]

Marshall stressed the importance of the peacetime innovation that the US had effected since the early 1990s—a luxury afforded by the Soviet decline. He envisioned the challenges to come, but during the relatively peaceful years that followed, he called for undertaking a more active search for and experimentation with new doctrines. Addressing the implications for strategic management, the assessment called for the following specific actions: to implement new concepts of operations and organizations through changes in educational programs and changes in acquisition and to create new promotion paths to train and to promote officers with appropriate skills and expertise. The memorandum also offered the SoD and the CJCS an opportunity to establish bodies and organizations that would develop knowledge concerning the military revolution. The memorandum recommended encouraging the services to entrust their very best people with the intellectual task of thinking about the future of warfare.[69] After conducting several historical studies sponsored by ONA,[70] Allan Millett and Williamson Murray concluded that "military institutions that developed organizational cultures where serious learning, study, and intellectual honesty lay at heart of preparation of officers for war, were those best prepared for the challenges that they confronted on the battlefield."[71] *The MTR Preliminary Assessment* became the intellectual starting point for the future US Defense Transformation.[72] Marshall and his proponents succeeded not only in intellectually defending their vision but in actually implementing the notion of the RMA across the US defense community.[73] The evaluation was circulated in the US defense community, initiating the most comprehensive reforms in the DoD since the Vietnam War.[74] A year after the publication of Marshall's legendary memorandum, there were five task forces exploring the RMA and its consequences.[75] From the mid-1990s on, the term *RMA* established itself among specialists as an authoritative frame of reference within which the debate over the future of war unfolded.[76]

The Role of ONA and of Andrew W. Marshall

There were individual analysts (most notably William Odom) who had been able to accurately assess Soviet MTR writings.[77] However, as a body, ONA outperformed all other segments of the US intelligence and defense community in this particular realm.[78] Marshall and Wohlstetter were the first Americans to introduce the idea of discontinuities in the methods of fighting into American military thought and the defense community. According to Richard Perle, the 2003 Operation Iraqi Freedom (OIF) was fought exactly along their vision of future war.[79] Although the most obscure, relative to other institutions of the American defense establishment, and most of the time below the radar, ONA significantly influenced US defense policy for several decades and was probably the most important institution in shaping US military thought from the late 1980s.[80] Because of the central role it played in the context of the RMA, this section will discuss briefly the intellectual history of the Pentagon's Office of Net Assessment.

Dissatisfied with the quality of long-term strategic thinking, in 1971 President Nixon established a unit of experts that would integrate intelligence and defense analysis on national security threats, which would report directly to National Security Advisor Henry Kissinger. The group was headed by Andrew Marshall, a graduate in economics from the University of Chicago, who had joined RAND in 1949 and who by 1969 had succeeded James Schlesinger as RAND's director of strategic studies. Marshall worked on a vast number of strategic defense and intelligence issues, together with the leading American defense intellectuals, and was a close friend and colleague of Albert and Roberta Wohlstetter. In 1973, when the unit was moved to the Pentagon, Andrew Marshall was named director of the Office of Net Assessment in the Office of the Secretary of Defense. The unit focused on analyzing competitive strategic environments between the US and the Soviet Union. To produce assessments of the nature of Soviet capabilities and of subsequent American countermeasures, it synthesized all available political, military, economic, and sociocultural data from within and outside of the defense and intelligence community. Frequently, ONA assessments brought up scenarios and issues that no other body had even considered, let alone assessed. Until now, Andrew Marshall's unit has advised twelve secretaries of defense from eight presidential administrations. The few available publications that discuss his work define Andrew Marshall as a "leading defense strategist," "unconventional long-term thinker," "one step ahead of the competition," and the experts of his office as the "Delphic Oracles" of the American defense system. The credo of this "bastion of futuristic brainstorming" was "to think out of the Pentagon box" on "important but overlooked" national security issues. Since the mid-1970s, this small office consisting of a dozen experts had become an intellectual center of gravity of the American defense bureaucracy.[81] Marshall is "the only prominent strategist from RAND's golden age still in government service."[82]

The idea of net assessment was dictated by the necessity to link defense policies with the anticipated reactions of opponents.[83] As an analytical tool, net assessment was neither art nor science and was distinct from system analysis, game theory, operations research, or strategic planning.[84] These formal analytical frameworks made predictions on the basis of players' rationality, assuming that strategic adversaries are following cost-benefit optimization rules. Net assessment parted ways with this conventional analytical wisdom because it lacked the tools to consider the "soft," immeasurable asymmetries between the competitors. It was distinguished by the consideration of unexpected outcomes that emerge from unforeseen and unappreciated factors, some of which might initially seem totally irrelevant.[85] As Rosen put it, "net assessments sought to avoid the natural tendency to assume the enemy would behave as we would, were we in his position, or that our forces will engage like forces on the enemy side."[86] For example, when analyzing an adversary, ONA considered sociocultural-bureaucratic operating patterns more than officially approved strategies. Net assessment integrated the strategies of the "biased" adversaries in a single space and simulated a competitive dynamic between them. On this basis ONA specialists generated their long-term forecasting.[87]

The memoirs of ONA veterans indicate that Mr. Marshall's office had developed its peculiar style of thinking, distinguished from the one practiced by the other segments of the American defense establishment.[88] ONA experts cared no less about imagination than about methodology.[89] Although acknowledging that futurology is often wrong, Andrew Marshall's motto was that "one can get many things right just by thinking about them a little bit." According to Paul Bracken, an ONA veteran, it was "a way of tackling problems from certain distinctive perspectives."[90] Andrew Marshall is considered to be the central figure to apply the net assessment methodology to security affairs.[91] The multifaceted ONA net assessments were both regional and functional.[92] To assess the military, political, and economic relationship between the US and the USSR meant for Andrew Marshall, first and foremost, reading and learning about what the Soviets were saying to themselves.[93] In 1982 he wrote:

> A major component of any assessment of the strategic balance should be our best approximation of a Soviet-style assessment of the strategic balance. This must not be the standard U.S. calculations done with slightly different assumptions ... rather it should be an assessment structured as the Soviets would structure it, using those scenarios they see as most likely and their criteria and ways of measuring outcomes. This is not just a point of logical nicety, since ... the Soviet calculations are likely to make different assumptions about scenarios and objectives, focus attention upon different variables, include both long-range and theater forces, and may at the technical assessment level, perform different calculations ... use different measures of effectiveness, and perhaps use different assessment processes and methods. The result is that Soviet assessments may substantially differ from American assessments.[94]

The eclectic, synthetic grasp of reality that Marshall established and culti-vated at the ONA impelled it to go out of its way to extend its reach—outside of the intelligence community as well—to gather empirical evidence from all the sources it could tap. During the 1970s, Marshall hired Sovietologists to carry out an interview project with Soviet émigrés. The CIA and DIA gener-ally ignored these people on the assumption that they were too biased to be useful sources of information. Marshall, by contrast, assumed they could be in-valuable sources of insight and went out of his way to see what they had to offer.[95] He was behind the funding of virtually all nongovernmental research carried out in the US on this topic.[96] During the last decades of the Cold War, Marshall was perhaps the most consistent, and retrospectively prescient, critic of CIA estimates of Soviet defense.[97] Marshall based his understanding of the Soviet MTR literature on reading for himself what the Soviets were actually saying as opposed to accepting official intelligence summaries. He was one of the first researchers to notice that American expectations about Soviet strate-gic behavior did not coincide with the projections of nuclear-force develop-ments of the 1960s. In 1972, Marshall criticized the American approach to nuclear strategy, arguing that the US "had followed a rich nation's strategy of attempting to compete with the Soviet Union in all areas of technology." From his time at RAND, he had become convinced that the Soviet military think-ing was fundamentally different from that of the Americans. In the context of net assessment work on the US–Soviet strategic–nuclear and NATO–Warsaw Pact balances, he had directed extensive research into Soviet military theory, measures of effectiveness, and assessment methodologies.[98]

When Soviet MTR writings were regarded by most observers as signals of new advances in Soviet military technology, Andrew Marshall saw a com-pletely different picture.[99] As Marshall himself later recalled, the Soviets were the first to bring to his attention the idea of revolution diagnostics. The diag-nostics entailed examining entire decades, to identify the emergence of new forms of warfare that were destined to dominate military conduct. It seemed reasonable to him to refer to the discontinuity discussed by the Russians as a revolution, and to diagnose it as early as possible. In Marshall's view, when the Soviets declared that the world had entered into a new period of Military-Technical Revolution, in fact "they were consciously experiencing a change in the nature of war. Usually, when one is in the middle of it, he is least aware of it. However, the earlier the military acknowledges the emergence of the change in the military regime, the more efficient defense management it will generate."

Marshall was, for the most part, in agreement with the Soviet methodological-theoretical approach to the nature of war in general as well as with the content of their analyses regarding the current MTR. Marshall explained that in a number of previous military innovations, for example, as with the Germans and the British in the interwar period, people in the military were not consciously thinking in terms of a revolution. He deliberately wanted to introduce practice, which was so natural for the Soviet military intellectuals, to diagnose the periods of significant change in the nature of war. According to his own account, the Soviet theoretical writings of the 1970s first brought to his attention the historical period of the twenties and thirties as an exemplar of such a change. In the years to follow, the proponents of the RMA and of Marshall's ideas would cite this period as a frame of reference for the discussion about the emerging discontinuity. He wanted American military theoreticians to self-consciously pursue and experience the emerging Revolution in Military Affairs.[100] This approach was in keeping with the traditional focus of ONA experts on strategic diagnosis, and not prescription. Marshall believed that getting questions right was more important at the initial stage than trying to get the right answers.[101]

However, ONA had as many bureaucratic weaknesses as it had intellectual strengths. It was a small advisory body with its own budget for independent research but had neither command authority nor budget allocation responsibility within the military or defense system. A small number of ONA experts, the long-term deductive thinkers of the American defense establishment, were engaged in activities that were being performed in the USSR by thousands of people in military academies and in the GS. Consequently, relative to their Soviet counterparts, they were less influential organizationally and could cover fewer topics. ONA did not inquire into the nature of warfare per se; it concentrated on Russian perceptions of it and generated its insights from there. Since ONA's assessments had not been instituted as an integral part of any military or defense program, its bureaucratic influence varied with the changing administrations. Several secretaries of defense made use of ONA as an advisor, while others preferred to draw on other sources of strategic analysis, placing ONA in the background. As an in-house Pentagon think tank, its authority never went beyond giving advice to the secretary of defense. ONA maintained a "polite distance" from the rest of the DoD and JCS, and did not regard its assessments as prescriptive. In ONA's view, it was up to someone else to act on its insights.[102] It never had the organizational tools, nor did it regard channeling its insights into immediate defense programs to be a part of its mandate. In contrast to the Soviet case, in which senior military leadership with command authority was involved in generating long-term forecasts, ONA's assessments were not necessarily translated into organizational or conceptual transformations.

The Implementation Stage of the American RMA

Following the overwhelming victory of coalition forces in Operation Desert Storm, a good deal of discussion took place as to whether the world had witnessed an RMA.[103] The RMA discourse of the 1990s became an "organizing principle of US defense modernization discussions."[104] According to Barry Watts, this innovative effort to think about how warfare might change in the years ahead spawned official work within the DoD as well as a wave of literature from the think-tank industry.[105] Throughout the early 1990s, the RMA thesis shaped the cognitive landscape of American military thought. According to Robert Tomes, ONA developed an intellectual climate that favored "thinking about changes in military affairs."[106] Marshall's office had taken the lead in financing studies on the history of military innovation in the interwar period—innovation that led to developments such as carrier strike aviation, amphibious warfare, and Blitzkrieg—and in sponsoring its own war games, conferences, and other RMA studies.[107] By 1995 a heady vision associated with the evolving RMA thesis aroused tremendous excitement among American defense planners.[108] The 1997 QDR "acknowledged the existence of the RMA" and urged to transform the armed forces accordingly.[109] From Desert Storm to Operation Iraqi Freedom, the notion of the RMA transformed American military forces tremendously.[110] Current American defense establishment "transformation" is, according to Colin Gray, a "logical and practical consequence of RMA."[111]

Although ONA experts tried to focus the professional attention of the US defense community on the symbiotic relationship between technology, concepts, and organizational structures, "techno-euphoria" thrived after the first Gulf War. "From the outset," argues Watts, "Marshall and Krepinevich were clear in their own minds that operational concepts and organizational adaptations were, if anything, more important than either new technology or getting it fielded in a significant number of systems."[112] An emphasis on very long time frames, the development of appropriate doctrine, organization, and practical operations, as opposed to a focus on technology alone, distinguishes Marshall's approach to a future RMA from most of the US defense community and from American cultural predispositions in general. Andrew Marshall regarded the changes in fighting that occurred during the interwar years of 1918–39 as combined-systems revolutions. In these kinds of innovations, while technological advances were necessary, the underpinning was a symbiosis between systems, doctrine, and organizational developments.[113]

However, as Gray precisely states, "despite the sophisticated and originally fairly tentative, essentially speculative view of Marshall and OSD Net Assessment, once the RMA idea became general property it was captured by a profoundly technological view of the revolution that seemed to beckon the Armed Forces into a new golden age of enhanced effectiveness. This techno-philia was to be expected, given America's technological strengths, its military culture, and its preferred way of war."[114] Lock-Pullan believes the technological lessons overshadowed the conceptual, social, and cultural ones. The list of examples he provides to support his argument is compelling.[115] According to Watts, "because the American military establishment has been so resistant to making the intellectual effort to come to grips with the challenges of this period, the Pentagon has probably not yet gotten even a third of the way down the road to mastering the changes in war's conduct foreseen in Marshall's 1992 RMA assessment."[116] Flawed thinking about the impact of technology on the character of future war occurred not only at the stage of the paradigmatic change. H. R. McMaster has shown how the US military frequently failed to understand the implications of the RMA. The superficial thinking that accompanied the uncritical embracing of the RMA corrupted American strategic and operational thought in subsequent decades.[117]

PART 2: AMERICAN STRATEGIC CULTURE

Cultural Characteristics and Cognitive Style

The normative image of American culture is individualistic. American society tends to emphasize goals and individual accomplishments rather than the rationale pursued by a group. Individuals are motivated by self-interest and personal values. The US is a performance- and achievement-oriented society concerned with assertiveness, heroism, and material wealth as a sign of success. One is expected to be an ambitious and competitive achiever; the quick and successful are respected and admired. Social hierarchy does not impose much on the individual; social bonds are flexible and fragile. American communication style is an example of the low-context culture. It is open, dramatic, precise, and explicit, ensuring that the listener/reader receives the message exactly as it was conveyed. During negotiations, ceremonial "beating around the bush" is not considered a virtue, while getting "straight to the point" is highly valued in the low-context American entrepreneurial culture. Using Edward Hall's terminology, scholars define Americans as a culture with a strong inclination to monochronism—considering issues independently from one another and contemplating one thing at a time.[118]

Assuming a correlation between the sociocultural structure of the society and its cognitive style, cultural psychologists recognize a strong predisposition of the American culture to the "analytical-logical" type of reasoning.[119] The so-called field or context independence inclines the subjects to focus on the particular object, to detach it from its context, and assign it to categories. Individuals with an analytical-logical grasp of reality pay attention to the salient object independent from the context in which it is embedded. This style of thought uses formal logic to explain and to predict an object's behavior. An observed phenomenon is dissected into pieces, which are linked in causal chains and categorized into universal criteria.[120] Sociologists argue that Americans assume that rational thinking is based on objective reality, where measurable results can be attained.[121] This manner of thought is well suited to the general American functional approach that emphasizes solving problems and accomplishing tasks. American mental formations favor procedural knowledge, which focuses on how to get things done, in contrast to descriptions of the way things are. This way of thinking is believed to be rational and efficient.[122]

This practical orientation of thought is consistent with a mental preoccupation with causality. Relative to other cultures, the idea of "natural happening" and "occurrence" is less acceptable to the American intellectual tradition. In contrast to holistic thinkers, this orientation creates a tendency to put data together in linear cause-and-effect chains along a single dimension. It is based inter alia on the optimistic belief that there is an objective essence that can be reached through linear processes of discovery. American experience employs psychology, game theory, and mathematics as dominant analytical approaches to channel human thinking and judgments into applications. It is a tendency that exists in symbiosis with the cult of the use of probability in risk analyses, which have penetrated through American culture.[123] Systems analysis prospered as an analytical instrument in military affairs under Robert McNamara. During the 1960s, progress in the Vietnam War was even measured by "body counts and track kills."[124] Practicality and functional orientation of thought aims it toward the nearest future, which appears in American think-ing in the form of anticipated consequences of actions.[125] When trying to en-vision the future and to make effective forecasting, logical-analytical cogni-tive style predisposes American mental formation to focalism—a tendency to focus attention narrowly on the upcoming target event and not enough on the consequences of other future events. Among other things, it is another symp-tom of "field independency," when little attention is paid to the overall frame-work in which rational and effective actions take place.[126]

Anthropologists argue that American social institutions and educational organizations emphasize systematic and analytical thinking and provide a less favorable environment for holistic conceptual reasoning.[127] The subjects of the American cultural climate have been described as manifesting a "logical-analytical" approach across the large spectrum of human activity.[128] The American personality is distinguished to a significant extent by practicality, which acknowledges the superiority of praxis over idealized concepts or ab-stract virtues. As Andrew May argues, American science and technology were historically produced in the most rational, pragmatic, and entrepreneurial fash-ion. The aim was not to create abstract, fundamental knowledge, but to create new, applicable technologies that could be widely sold or used to generate profits.[129] This predisposes American conduct toward empiricism, a theory of knowledge that emphasizes the role of experience and those aspects of sci-entific inquiry that take shape through deliberate experimental arrangements, in contrast to the a priori abstract knowledge that is independent of experi-ence and experiment. Donnelly notes the strong predisposition of the Ameri-can military tradition to value practice at the expense of theory.[130]

On balance, the mainstream of American thinking style has historically manifested a strong inclination toward inductive reasoning—it derives prin-ciples from analysis of data and generalizes from the facts to the concept.[131] Robert Bathurst shows that in the military realm and in international negoti-ations, the United States began analyzing situations from fragments. Keeping with its low-context orientation, its practical approach concentrated on the most rudimentary aspects of the problem and sought the most optimal mechanical solution to it.[132] In discussing American negotiation style and legislation, Cogan also describes an American proclivity toward inductive reasoning. He dem-onstrates how this empirically based, inductive style of reasoning contrasts with the Cartesian mode of deduction. While the deductive approach moves from "top-down," first defining principles and then deducing the context, the

American approach is "bottom-up," moving from the particular to the general. This orientation also exists in American legal practice.[133] Comparative studies of political reasoning suggest that induction and empiricism are characteristic of Anglo-American societies, as opposed to the holistic-dialectical thinking of several other cultures.[134] This intellectual climate fostered America's cultural approach to innovation. In his most thorough and original analysis of US strategic culture, Brice F. Harris argued that "the story of scientific discovery in America is less the story of scientific research in its *creative* form— which is to say research in the tradition of Aristotle, Newton or Einstein—than it is one of *applied* research."[135] As was established above, creative holistic-dialectical thinking style has a greater aptitude than its logical-analytical alternative to grasp paradigmatic changes in military affairs.

American Culture of War: Swift Annihilation and Attrition by Firepower

The national mission of conquering an entire continent, together with the nation's prolonged frontier experience, left its mark on American strategic culture.[136] The United States developed into a country of unusual dimensions, and the scale of its resources has influenced the national security enterprises it has undertaken.[137] As American society grew in size and wealth, it also accumulated military power, with no apparent economic or demographic limits. Restrictions on American power were not natural, but rather determined by political and strategic considerations.[138] Almost two hundred years have passed since the United States faced an enemy with a larger gross national product than its own. American productive capacity, translated into overwhelming material superiority, has played a critical role in the nation's military successes. Its dominance in numerous industrial and technological sectors, in skilled manufacturing, and in the ability to increase production capacity created basic military advantages: a large defense budget, a significant pool of machines for fighting a war, and educated manpower capable of operating them.[139] Given the abundant material resources, troops' equipment, and excellent managerial expertise, the United States relied less on perfectly planned and executed strategies to win.[140]

Self-efficacy dictates a strategy to shift the conflict into those arenas where one enjoys an inherent advantage over one's enemy.[141] The strategy of attrition and annihilating the enemy with firepower was the best way to transform the nation's material superiority into battlefield effectiveness. The translation of enormous resources into firepower, technology, and logistical ability and a consequent inclination for direct attack date back to the military experience of the American Civil War. This "annihilation by fire" approach has been largely successful throughout American military history.[142] In illustrating this tendency in the country's strategic culture, Cohen points to two outstanding characteristics of American conduct during the Second World War: a preference for massing a vast array of men and machines, and a preference for direct assaults.[143] According to John Ellis, on the operational level, US generals relied on material superiority, firepower, and overwhelming force rather than on creative maneuvers that would threaten the enemy and force them to surrender.[144] Referring to the American preference for mechanical and industrial solutions, some argue that the United States has often waged logistic, rather than strategic, wars.[145]

Discussing American strategic culture, Mahnken defines this preference for an overwhelming blow as taking a "direct approach to strategy over indirect." In his discussion of American strategic culture, he dubs this phenomenon "an industrial approach to war."[146] Echoing this claim, Chester Wilmot argues that the Americans have adhered to the theory that if a military machine was big enough, it could be driven wherever they wanted to go.[147] The conflicts in Korea and Vietnam provide further examples of the military doctrine of annihilation and a resource-based approach to warfare.[148] Capitalizing on this industrial approach, the US has often outproduced its enemies in the amount of military power that it is able to generate.[149] Criticizing this formulation, other scholars have insisted that the American armed forces have pursued a much wider range of strategies beyond pure attrition or annihilation. American military tradition, they argue, is also rich in fighting small wars and insurgencies and in developing excellence in deterrence strategies.[150] However, Mahnken has claimed, even in these cases, a preference for attrition and annihilation "stands up remarkably well as a portrayal of American military strategic culture and the aspirations of the US military."[151] It is most likely for this reason that US strategic culture, which seeks decisive battles, ultimate victories, and measurable national security outcomes, is less at home with stability and support missions, on which swift annihilation by massive firepower is less relevant.[152]

A-strategic Thinking

Longstanding American superiority in resources translated into a traditionally low incentive to engage in patient strategic considerations and in thorough operational calculations.[153] Scholars agree that the materially wealthy United States has, throughout its military history, preferred an approach to war based on annihilation and attrition by means of technology and fire-power over a style of fighting resting on maneuverability or on strategic thoroughness.[154] The American military sought to take the war to the enemy as rapidly and as destructively as the machinery of industrial age warfare permitted, while maneuver was considered to be simply the means to impose firepower on the opposing force.[155] It almost took for granted that it would be able to mass forces and firepower whenever and wherever it desired.[156] This industrial approach to warfare accounts, according to some scholars, for the relative disfavor with which traditional military theory is regarded.[157] Lock-Pullan notes that the United States did not historically develop "excellence in strategy and military thought because it did not have to."[158] Scholars report the strong predisposition of the American military tradition to value practice at the expense of theory.[159] Although a professional military education of the US officers' corps was strongly emphasized, Murray argues that American strategic culture frequently tended to be anti-intellectual and antihistorical. According to him, the US senior military leadership's "overall attitude at best appears to be that education is a luxury for the American military rather than a necessity."[160] According to Gray, this neglect of a professional military education at the top results, in part, in a tendency to think a-strategically.[161] The philosophy of a continuous and profound professional military education was simply not that important an attribute to American military culture. Intellectual curiosity in military science never became a criterion for promotion.[162]

The above observations also reflect on the American approach to developing professional theoretical knowledge about the nature of war. No theoretical approach for the organized study of war in all its aspects (the impact of social, economic, political, and technological phenomena on the methods of fighting) was ever formulated in the West.[163] Fundamental studies of war and predictions about its future obviously did take place in the US. However, when scholars compared them to those done in the USSR, they found the former to be fragmented, not integrated, uncoordinated, and rarely linked directly to the development of the state's military machine.[164] Edward Luttwak, in an essay written in 1981 in *International Security*, pointed out that despite the long-standing recognition of an operational level of warfare in classical military literature, there was no adequate term for this in Anglo-Saxon military thought.[165] John Erickson and Raymond Garthoff have concurred that the term *operational art* was foreign to Western military thinking.[166] This was a serious conceptual shortcoming, since it is generally on this level that paradigmatic changes in the nature of warfare are debated. Strategic and tactical implications are an outgrowth of the initial insights produced in the milieu of operational art.[167] The American disinclination to invest in operational thinking comes as no surprise. The idea of "collapsing the enemy" by operational maneuver rather than simply annihilating it by firepower conceptually diverged from the established American strategic tradition.[168]

It was only in the wake of the Vietnam War that ALB concepts began to emphasize warfare maneuverability and the necessity to develop theoretical knowledge on the operational level. It was not until 1986 that the US army reoriented from battles of integral annihilation in favor of a more dynamic and complex understanding of war,[169] and officially recognized operational art as an integral part of US military thought.[170] In general, the American strategic community developed an approach to strategy that accounts less fully "for the range of strategic dimensions" than does that of Russia or China.[171]

Optimistic and Engineering Approach to Security

The belief of the founding fathers that America represented a "new beginning" contributed to a national identity based on liberal, democratic, Protestant, and capitalistic principles. Individual freedoms, pragmatism, and rationalism formed the cornerstones of the new society.[172] The capitalist economy, liberal political structures, and a strong spirit of exploration produced a belief that as nature could and should be understood, potentially almost any problem can be solved. Optimistic entrepreneurship became a value in all fields of American social activity and created a society based on notions of efficacy, rationalism, and pragmatism. Compounded by repeated success, it produced a romantic engineering creed that viewed social and security problems as essentially mechanical in nature and, consequently, consistent with the logic of man-made machines.[173]

American history is rife with "miraculous" achievements, typically in the face of challenging geography. Conquering the wilderness bred a frontier pragmatism that was translated into an engineering, problem-solving ethos. This approach often regards political conditions as a set of problems,[174] and pushes strategists, influenced by engineering, to "attempt the impossible."[175] A belief evolved in popular culture that problems could always be solved.[176] As a society whose Declaration of Independence affirmed the "pursuit of happiness" as the natural right of every citizen, the Americans tended to take a proactive

approach, viewing sources of unease and discomfort as "engineering problems."[177] The political challenges posed by the American Indians or menacing
European or Asian empires were transformed by the United States into military problems that could be resolved definitively by means of machine warfare.[178] The absence of national-level security disasters reinforced optimism as
an American national philosophy.[179] Such a strategic culture is more at home
with administration than with the art of diplomacy or strategy.[180] It is inclined toward reductionist methods of problem solving, by minimizing the
complications created by culture, time, and distance.[181]

Similar engineering positivism is manifested in American military thought.
Though Carl von Clausewitz might be considered the father of the American
approach to civil-military relations and strategy, many claim that the true mentor of US military thinking is Antoine Jomini. He wrote about war as an art,
but his quest for reducing complexity to a few apparently simple principles
has characterized the cultural preference of American military thought for
simplicity over complexity. Armed with the Jominian belief in the effectiveness and power of basic axioms, American practicality sought to reduce strategic problems to equations. Historically, American strategists have assumed
that they could calculate the answers to the issues of deterrence and war. The
country's domestic history encouraged the belief that American know-how
would inevitably find a solution to any problem.[182] This tendency is reinforced
by an American fascination with technology that dictates, drives, and orga-
nizes the managerial mind-set in military affairs.[183]

American Time Orientation—"Present and Immediate Future"

Scholars of American strategic culture describe how the need for immediate
action, the rapid resolution of problems, and achieving results went hand in
hand with a strong American time orientation toward the present instead of
the past or distant future.[184] The practicality of American thinking "condemns
the irrational past" and directs it toward the immediate future, making the
orientation more functional than that in other societies, where the future is
measured in decades or generations. American time, argues Edward Hall, is
linear.[185] The future appears in American thinking in the form of anticipated
consequences of actions.[186]

American culture usually considers the newest to be the best. This inclination is clearly reflected in the US military's approach to weapons acquisition
policy. While Soviet weapons research, development, and procurement were

driven by consumer requirements, the Western armed forces often procured what industries produced and sold. In the West it was possible for a weapons system to be procured because it represented state-of-the-art technology, and not necessarily because its use was prescribed by the doctrine.[187] The fascination with novelty and rapid transformations predisposes American society to accept change more readily than do other cultures. However, as Downey and Metz have noted, with little attention paid to the past, the tendency is to look ahead—not to the distant future, but more to the demanding present time.[188] Although US strategic planning has not always focused solely on the here and now,[189] observers characterize it as generally averse to an extended strategic outlook and more comfortable with near-term crisis management than with long-term strategy planning.[190] As Murray put it, referring mostly to Vietnam, "the American nation's worst defeat resulted largely from a military and civilian leadership that prized modern technology over the lessons of the past."[191] This lack of historical and cultural curiosity frequently results in a situation in which the enemy of the US understands the Americans far more coherently and effectively than the Americans understand him.[192]

Democratic Tradition, Bottom-Up Organization, and the Role of the JCS

As a social-organizational phenomenon, the JCS manifested the American strategic culture just as the Soviet GS was consistent with the Russian cultural characteristics. The American political modus operandi prevented the concentration of an ultimate authority in a single military organization. Consolidating bureaucratic power in one central place (i.e., in the hands of the JCS chairman) would have contradicted the American democratic tradition of checks and balances. In keeping with the liberal tradition of American society, authority was delegated down to the services. Consistent with an entrepreneurial culture, the competition between services was expected to be beneficial and to serve as an impetus for innovative initiatives.[193] As a result, one of the most significant bodies of the American military system, the JCS, was also one of the most controversial. Although the JCS was designated as the principal military advisory body to the civilian leadership, the chairman lacked the statutory mandate for independent long-term recommendations. His advice centered more on budget allocations and less on long-term strategy or development of American military power. The JCS was, for the most part, disconnected from the operational realm, rarely held command responsibility of its own, and as a rule, delegated considerable authority, including doctrinal development, to the services.[194] De facto, the services, and not the JCS or the Department of Defense, were the most powerful institutions of American national security.[195]

The establishment and subsequent functioning of the JCS was a distinct manifestation of American military parochialism. Its members faced a constant built-in dilemma, between representing the interests of their respective services and thinking jointly and broadly about the nature of the armed forces in an existing or emerging security environment. It was the former that prevailed. Rather than an elite military organization that concentrated the finest professional capital, the selection process produced narrowly focused, combat-oriented line officers, committed to the parochial interests of their services. The officers were selected late in their careers and were not formally educated for duty in the JCS. In striking contrast to the Soviet GS, the JCS by no means consisted of the intellectual crème de la crème of the American military.[196] Strategic and long-term defense planning were weakly institutionalized in the JCS. It lacked the powerful cadres required to produce effective cross-service vision and advice that was capable of affecting the long-term development of the US military. By definition, the chairman was a budgetary manager and occasional operational planner, but not a deductive thinker about the nature of war. He was neither a doctrinal luminary nor an initiator of long-term strategic decisions. The JCS remained a captive of the services and lacked the intellectual mechanisms to generate broad, cross-cutting, long-term recommendations. The institutionalized conceptual centers of gravity, such as Training and Doctrine Command (TRADOC), were diffused among the services, which initiated most American military innovations.[197] Lacking strategic guidance, the services' innovations were often piecemeal, inconsistent, and sectarian, and they rarely expanded beyond the operational level. "Each branch developed its distinctive strategic paradigm," and the JCS rarely offered conceptual alternatives to the views developed in the services.[198] As a rule, American military innovated bottom-up, from the services to the leadership.[199]

No single institution existed in the American military that possessed a synthetic grasp of the security environment. Given the structure of the JCS, there was no institution capable of systematically thinking through the discontinuities in military affairs along the entire spectrum of their implications for the services. Without that perspective, it was virtually impossible to analyze the impact of the scientific-technological changes on the nature of war-fare in general and on the doctrine and organization of the American military forces in particular. The state of professional periodicals serves as a case in point. Following the 1986 military reform, the JCS established its own profes-sional publication, *Joint Forces Quarterly,* for the dissemination of knowl-edge among senior military professionals. This vanguard of American military thought was established only in 1993. In contrast, the professional publications of the American military services had been established several decades earlier than the quarterly of the JCS. For the sake of comparison, the Soviet GS had established its periodical *Voennaia Mysl'* (Military Thought) in the late 1920s. The titles of the journals also manifest which raison d'être their founders at-tributed to them. Murray, in discussing the relatively insignificant attention paid to doctrinal conceptualizations and theory development within the frame-work of American military culture, argues that the Joint Staff never consti-tuted the intellectual center of gravity of the US military forces.[200] Through the years, the above-mentioned weaknesses of the JCS system were observed and noted by several American defense intellectuals.[201]

Technological Romanticism in Military Affairs

The strong bias toward techno-centric warfare is an essential component of American strategic behavior. Scholars do not condemn prudent exploitation of the technological dimension of war, but rather its misuse and overreliance on machines.[202] According to Mahnken, "technological optimism has histori-cally animated US defense planning"; "no other nation has placed greater emphasis upon the role of technology in planning and waging war."[203] Reli-ance on new technology is a corollary of the predisposition to solve problems quickly and in simple, direct terms.[204] Initially, America's romance with machin-ery, particularly with mechanical means of transportation, was a result of the need to conquer the wilderness. Population density on the frontier, together with an acute shortage of skilled artisans, obliged Americans to invent substi-tutes for human skill and muscle.[205] The new society responded to this short-age by ingeniously embracing machines and taking the lead in the production of mechanical tools. Since the early nineteenth century the United States has been a land of technological marvels and has developed an extraordinary rate of technology dependency.[206]

The fascination with technology was not unique to the military; it character-ized the culture as a whole. In the popular narrative, technology is generally seen as bringing benefits. In contrast to Europe, American history has few examples of mobs destroying industrial machines. The capitalist economy demanded on-going innovation, while the character of the competition pushed scientists, technologists, and consumers closer together and produced unusually enthusiastic thinking with regard to technology.[207] The machine acquired symbolic meaning, and the liberal American tradition saw technology as an instrument for preserving the nation's immunity from war, rather than as new means for waging it.[208] One of the principal by-products of technology was a faith in technology.[209] American strategic culture viewed technology as a panacea in global affairs and sought ways to expand its scope and to apply technical solutions to strategic issues.[210] Weigley, in discussing the American way of war, argues that the pragmatic qualities of the American character have fostered a national penchant for seeking refuge from difficult problems of strategy in technology.[211] This predisposition to technicity—to the exaggerated significance of the technical—was characteristic of American defense policy makers, as well.[212] Many military historians regard the technology of warfare as one of the most important independent variables in the country's military thought.[213] Technological romanticism engendered visions of a mystical silver bullet promising decisive victory.[214]

The zeal for technology was further fueled by a desire to get more "bang for the buck" while minimizing American, if not enemy, casualties.[215] The desire to minimize human losses is another trait of American strategic culture. American society could not abide a high rate of casualties, and the military sought a style of fighting designed to minimize fatalities. Thus, it became American practice to send metal into battle in place of vulnerable flesh.[216] The preference to expend bombs and machines rather than personnel also led the United States to prefer to wreak destruction from a distance.[217] According to Cohen and others, these elements are mutually reinforcing. The armed forces opt for airpower, strategic bombing, stand-off strikes, overwhelming firepower, and high technology as a means to reduce the forces' vulnerability in military operations.[218] This technological enthusiasm varies across the distinct subcultures of the American military services.[219] The air force and navy were traditionally the most techno-friendly and techno-dependent. The army kept its distance from such techno-bias, and the marines valued technology the least. Being "boots-on-the-ground" services, the army and marines rely to a relatively greater extent on the human element than on machines and put the former at the center of warfare; hence, the saying that the air force and navy man the equipment, while the army and marine corps equip the man. How-

THE US REVOLUTION IN MILITARY AFFAIRS 87

ever, scholars agree that in spite of these differences, optimistic techno-centric romantic culture was ingrained in all four of the American military services and also served as a common denominator for civilian policy makers involved in military affairs.[220]

Within the defense establishment, debates about technology and budgets frequently usurped the place of strategy. The traditional orientation toward quick action and results, an attachment to things new and futuristic, and a disinclination to wage long wars were frequently in keeping with the almost instinctive reliance of American strategists on technology as a panacea in national security affairs.[221] This pragmatism can result in a technical approach to international security, and a conception of complicated issues as problems requiring engineering solutions.[222] American reliance on technology, according to Mahnken, was a poor but ubiquitous substitute for strategic thinking in international security.[223]

An Inclination to Ethnocentrism

The US has historically seen itself as an arbiter of morality, with a special moral-political mission in the world.[224] It has been argued that this vision, fueled by the isolationist tradition, has sometimes created an ethnocentric belief among Americans that they occupy the moral high ground, as well as their inclination to view the world primarily through the perspective of their own culture.[225] The early ideologies of American colonists were influenced by the Protestantism of the Puritan settlers, who believed that they were God's people—chosen to lead the other nations of the world. John Winthrop gave this notion metaphoric expression in his description of America as a "City upon a Hill."[226] The successful course of political and military history in the US has provided justification for its belief in its own optimism, a self-confident sense of superiority and invulnerability.[227] Americans' high estimation of themselves as a nation, including a collective narrative that emphasizes political and moral uniqueness, liberty, a divine mission, and a multidimensional sense of national greatness, has made it difficult for them to accept the beliefs, habits, and behaviors of foreign cultures.[228]

American history, at least up until the Vietnam War, was presented as an extremely positive narrative. Young colonies evolved into a power capable of carrying out the world's most important endeavors. This generated an extraordinary optimism regarding what could be achieved by the American way of war.[229] The early wars—the Seven Years' War (1756–63), the Revolutionary

War (1775–83), and the War of 1812—regardless of how they had begun, were victorious at a relatively small cost. The late American entry into both World Wars was followed by a steady march toward victory. Successful involvement in both wars is recounted with considerable ignorance, minimizing the role played by Britain, Russia, and France, and with a belief that the United States had ultimately won in both cases.[230] This was a narrative that perpetuated ethnocentricity and bolstered the existing strategic culture.[231] The overwhelmingly victorious historical experience kept Americans from examining counterproductive conduct that might undermine military effectiveness.[232]

Ethnocentrism is known to produce a phenomenon known as mirror imaging, a cognitive situation in which decision makers or intelligence analysts project their thought processes or their value system onto the subject of reference.[233] The tendency toward mirror imaging also comes from insufficient interest in the opponent's way of thinking. This "pathology" has been diagnosed in the American security and intelligence experience.[234] It primarily hampers the ability to properly identify and assess emerging foreign methods of warfare, and inclines to assess technical developments on the basis of the analyst's own technology.[235] Bathurst has reported on constant "mirror imaging" in the routine work of American intelligence officers analyzing the Soviet military doctrine and technological capabilities during the Cold War.[236] The adversary's practices are studied not only in order to understand the potential enemy, but also in order to learn alternative military art to emulate valuable ideas. In this regard, ethnocentrism can prove to be a serious obstacle. While the Soviet Army showed no reluctance to imitate and copy ideas from the US, this was not usually the case in reverse. Western nations, and the American military in particular, were less flexible in their attitudes, exhibiting a "not invented here" mentality.[237]

CONCLUSION: CULTURAL FACTORS AND
THE AMERICAN RMA

The cultivation of the technological seeds of the American RMA preceded the maturation of the conceptual ones. The US developed technology and weaponry for about a decade without realizing its revolutionary implications. Why did it take the US defense community close to a decade to acknowledge the accuracy of Soviet assumptions and to translate MTR theoretical postulates into a radical military reform? Several qualities of American strategic culture prevented its swift comprehension of the paradigmatic change in the nature of warfare discussed in the case under study.

The general American cultural predisposition to a logical-analytical cognitive style made the US military less receptive to the kind of reasoning needed for perceiving and comprehending dialectical transformations in military affairs. According to observations of cultural psychologists, American cognitive style was "field independent." It had a strong inclination to focus on the central object of the observed phenomenon and to pay little attention to the overall framework in which the action took place. During forecasting attempts, the logical-analytical style predisposed American mental formation to focalism—a tendency to focus attention narrowly on the upcoming event and to detach it from the overall implications. The pragmatic and practical American strategic personality predisposed to action, favored procedural knowledge that focused on how to get things done, and was inductive.

During the introduction of the PGM weaponry into the battlefield, mainstream American military experts concentrated on the *focal point*—mechanical application of the new technologies on the tactical level—and detached it from the *context*, that is, the implications of this new weaponry on the ways and means of waging operations. The concentration on the focal technologies at the expense of the broader contextual implications hindered the US military from the swift comprehension of the paradigmatic change in the nature of war. The US forecasting efforts were piecemeal and extrapolated from current capabilities, rather than trying to anticipate qualitative leaps in military regimes. The ALB innovation aimed to satisfy specific requirements related to a narrow, techno-tactical yet relevant set of operational threats. The US military for a long time saw in the stand-off PGMs no more than a perfect and immediate remedy to the Soviet echelonment doctrine. The US possessed only an intuitive understanding of the revolution that was about to occur, and was not consciously thinking in terms of a revolution. Not until Andrew Marshall and his colleagues from ONA introduced the notion of the RMA into the professional military discourse did the emerging discontinuity reach the consciousness of the American defense community. In keeping with the inductive approach to understanding reality, a paradigmatic change among the mainstream of the US military did not occur before the particular empirical experience (Gulf War) was observed and generalized.

Why did ONA experts reach better assessments than the rest of the American intelligence community on what the Soviets were thinking? As Gray noted in one of his publications, "a security community may have more than one strategic culture."[238] Indeed, several American leaders and analysts have challenged a techno-phile inclination of the US defense community. In Mahnken's words, these "latter-day Luddites decry the American military's seeming fascination with technology."[239] Taken over by Andrew W. Marshall ever since, the ONA experts succeeded in grasping this discontinuity because they did not embody—in fact had consciously suppressed—most of the traits of American strategic culture. The intellectual conduct of ONA was the striking exception to the rule, which proved the rule. Eclectic, holistic, and synthetic in nature, the thinking style and the intellectual atmosphere inside ONA were remarkably divergent from the logical-analytical approach of mainstream American strategic culture. ONA experts consciously stressed the importance of context dependence in the course of their analytical activity and sought to distance themselves from mainstream mechanical focalism. In contrast to the prevalent American cost-effectiveness approach, which was procedural and prescriptive, ONA was consciously committed to producing descriptive knowledge.[240] Better manifestation of ONA's deviation from logical-analytical thought and its inclination toward a holistic-dialectical thinking style can hardly be found. However, its influence was too limited to reorient and retool the fundamental American approach. ONA was too weak to struggle effectively against the general decontextualized thinking, technological romanticism, and intellectual myopia.

American strategic culture was less prepared institutionally and intellectually to think in terms of revolutions in military affairs. Institutionally, in keeping with decentralized liberal culture, relevant conceptual and organizational military innovations, such as ALB, originated in a bottom-up manner, from the services, and not top-down from the JCS or DoD. In keeping with the American cultural tendency to divide strategic problems into discrete parts in order to solve them, discerning the whole was frequently difficult. American JCS had no ethos of being a "brain of the military," and consequently strategic and long-term defense planning was weakly institutionalized there. The JCS lacked a powerful bureaucracy capable of producing an effective cross-service vision and advice that could affect the long-term development of US military power. The chairman was a budgetary manager and occasional operational planner, but not a deductive thinker about the nature of war. He remained a captive of the services' parochialism and lacked the intellectual capital to generate deep, cross-cutting, long-term observations.

Intellectually, the US military was unprepared for grasping the RMA. For generations, an integral battle of annihilation and enemy attrition by superior firepower had been an American way of war. This industrial approach to warfare accounts for the relative disfavor of the American theoretical military tradition. One implication was that the notion of the operational art as a theoretical concept was rejected by the US military tradition until the 1980s. The aim of "collapsing the enemy" by operational maneuver rather than simply annihilating it by firepower conceptually diverged from the established American strategic tradition. Because ideas about paradigmatic changes in the nature of war originate on the operational level of military thought, the lack of this intellectual layer was a serious obstacle that prevented thinking in terms of the RMA.

The traditional orientation toward quick action and results, an attachment to things new and futuristic, and a disinclination to wage a long war resulted in almost instinctive reliance of American strategists on technology as a panacea in national security affairs. An optimistic and engineering approach to security, an industrial approach to warfare, annihilation and attrition by firepower, the positive role of machines in the American cultural narrative, and the desire for cost-effective firepower, while minimizing casualties, made the US probably the most techno-centric military in the world. In this atmosphere, functional and mostly tactical application of advanced technologies took center stage. With certain variations, techno-euphoria was deeply ingrained in all four military services. During the 1970s, this technological romanticism disinclined the defense establishment to perceive the broader impact of this technological breakthrough upon the nature of war and to make a quantum leap in the sphere of military thought. American thinking appeared to focus more on how new technologies could be used to enhance performance of existing missions. The PGMs were seen as just another, albeit significant, force multiplier in the military arsenal. Notwithstanding ONA's intent to focus the professional attention of the US defense community on the symbiotic relationship among technology, concepts, and organizational structures, techno-euphoria blossomed once again during the implementation stage of the American RMA in the late 1990s.

Historically, ethnocentricity was a considerable factor in American strategic culture. The US saw itself as an arbiter of morality, with a special moral-political mission in the world. This vision inclined to view the world primarily through the perspective of its own culture. Ethnocentricity increased the likelihood of such analytical pathologies as "mirror imaging," in which foreign security developments were measured by American standards. This unmotivated analytical bias of the US analysts made them less receptive to certain military innovations from abroad, since they did not correspond to the common wisdom of the American defense establishment. In keeping with this cultural trait, in their evaluations of the Soviet MTR, American experts projected their own perceptions. They measured the Soviets by the standards of the US military and on the basis of American technology. Until the ONA assessment, the US defense community had failed to grasp the essence of the Soviet MTR developments and avoided accepting controversial futuristic conclusions offered by the Russians. Soviet writings about the revolutionary impact of the new weaponry were skeptically treated by the US experts as futuristic nonsense.

Ironically, Soviet theories ultimately provided a kind of a "mirror" for US strategists. By analyzing how American military power was reflected in Soviet eyes in the early 1980s, US strategists were able to realize during the early 1990s the value of the revolutionary treasure they held in their hands.

4 THE IMPACT OF CULTURAL FACTORS ON THE ISRAELI REVOLUTION IN MILITARY AFFAIRS

ALTHOUGH Israel was the first country to make extensive use of RMA technology, on the battlefields of Lebanon in 1982, it took more than a decade for the RMA debate to gather momentum in the Israeli Defense Forces (IDF). Israel lagged behind after the US and Russia in developing a conceptual framework that acknowledged the revolutionary implications of RMA. The purpose of this chapter is to trace the impact of Israeli strategic culture on the approach of the Israeli defense community to RMA.

This chapter comprises three parts. The first part concentrates on the intellectual history of RMA in the IDF. It discusses the development and use of the revolutionary capabilities on the battlefield and the subsequent attempts for military reforms; describes the acknowledgment of the RMA by the Israeli defense establishment; and refers to the transformation of military thought in the IDF. The second part analyzes the general sources and primary traits of Israeli strategic culture. It describes several fundamental Israeli cultural characteristics and touches concretely on such issues as the "Israeli way of war"; the quest to assume a qualitative edge; the role of technology; "siege mentality"; short-term planning and a tendency to improvise; the anti-intellectual tradition; and the role played by the GS. The conclusion integrates the previous two sections to discuss how the characteristics of Israeli strategic culture account for the conduct of the IDF with regard to the RMA. The chapter clarifies why, similar to the American case, the technological seeds of Israeli RMA preceded the conceptual ones.

PART ONE: ISRAELI RMA

The "Unconscious" RMA (1973–1991)

The story of the Israeli RMA starts on the battlefields in 1973, when doubts were raised about the credibility of the IDF's concept of operations. In Particular, two problems were highlighted: the low ability of the IDF to conduct combined arms operations, and the problem of the saturated battlefield. The latter resulted in high rates of attrition and significantly restricted operational maneuver, leaving no possibility for decisive breakthrough battle.[1] The efficacy of the traditional cult of the offensive began to appear reckless.[2] Hampered by numerous obstacles, the combined arms approach matured gradually in the decade that followed.[3] The issue of the saturated battlefield, however, created serious dissonance in Israeli military thought.

Inasmuch as static defense totally discredited itself in 1973, the notion of opting for *reactive defense* emerged. During the late 1970s, Saadia Amiel, scientific advisor to the minister of defense, argued that recent achievements in microelectronics and particularly on stand-off PGMs might be the most suitable means to this end. It was his view that while these advanced technologies were no panacea, they had the potential to enable on-time assessment of threats, real-time battlefield intelligence, swift target acquisition, efficient command and control, and precision fire strike.[4] His views found support among a select few[5] but were not adopted by the Israeli defense establishment. The conventional wisdom posited that a transformation to defensive doctrine would signal to the enemy that the IDF had no intention to invade their territory. This, it was argued, would decrease deterrence capability and would motivate the enemy to invest massively in building offensive measures. Israeli decision makers assumed that even if based on a remarkable PGM arsenal, defensive military doctrine would not be able to stop enemy waves. Intelligence, communication, and weaponry systems based on state-of-the-art technologies were procured, but neither doctrine nor force structure of the IDF underwent any significant change.[6]

The IDF's brilliantly orchestrated suppression of the Syrian air defenses in the Bekaa Valley demonstrated this dual approach, when advanced technologies were used in a revolutionary manner, without effecting a major change in the IDF's concept of operations. Lessons learned by the Israeli Air Force (IAF) after 1973 brought significant techno-tactical innovation—an exclusive development of the defense industries and the IAF. The IDF improved combined arms coordination, capabilities for dissemination of target intelligence, command and control, and precision fire.[7] In summer 1982, Syria introduced mobile and static SAMs into the Bekaa Valley, and created a daunting tactical situation for the IDF. Syrian deployment threatened Israeli air superiority and restricted the ability to conduct reconnaissance, interdiction, and close air support missions. Once the decision to engage the Syrian air-defense deployment was made, the IAF executed an attack based on the most advanced weaponry. In the first two hours of the battle, the IAF destroyed nineteen SAM batteries, and in the next few hours shot down more than twenty enemy MIGs, without losing even a single airplane. On about 1,500 square miles of airspace, the IDF executed paramount intelligence collection, target acquisition, command and control, and real-time data dissemination to precision-guided stand-off fire capabilities. Simultaneously the IDF jammed most of the Syrian ground communication and conducted electronic countermeasures against radars that were subsequently destroyed by anti-radar missiles. The knockout of the Syrian air and air-defense forces utilized the most revolutionary electronic and information warfare capabilities. The real-time battlefield picture was produced by the data fusion from the airborne surveillance radar, remotely piloted and unmanned flying vehicles, which monitored the area uninterruptedly. In the final stages of the battle, the IAF attacked Syrian armor with laser-guided missiles. The information was disseminated in real time, while sensors, command and control, and shooters operated as one orchestrated whole.[8] The Israeli cutting-edge weapons and technologies were so revolutionary that the superpowers had not yet produced, let alone deployed, analogous equipment.[9] The IAF had, in Soviet MTR terminology, operated as a gigantic, combined arms reconnaissance-strike complex. In American terms, Israeli conduct could be dubbed RMA-NCW-type battle par excellence, because it tailored intelligence and target acquisition systems, command and control elements, and precision stand-off fire into one integrated system and uninterruptedly implemented real-time sensor-to-shooter loops, for a relatively prolonged period of time.[10]

Despite the successful execution of this innovative approach, the experience did not produce any revolutionary transformation in the IDF's concept of operations.[11] Paradoxically, the Lebanon war led the IDF to favor quantity over examining conceptual, qualitative leaps forward.[12] Technologies and concepts of operation from summer 1982 were further developed in the IAF but became a sectarian, tactical innovation, with no significant impact on the concepts of operation of other corps, let alone GS. During most of the 1980s, the GS hardly had time to dedicate to the question of revolutionary implications of the 1982 IAF success.[13] Although pioneers in technology implementation, the IDF did not produce any significant leap forward in its vision of the future war and concept of operations.[14] Israel regarded the Bekaa battle as a successful synthesis of advanced technology and creative operational improvisation—the essence of its qualitative edge.[15] Thus, during most of the 1980s, state-of-the-art intelligence, communication, and precision fire systems were procured, without changing doctrine and organizational structure whatsoever. If new operation problems evolved, they were usually solved through additional force and sophisticated technological improvements.[16]

The question of doctrine effectiveness was raised in the 1980s, under the impact, among other reasons, of the *Wald Report*. This internal report criticized the IDF approach to force buildup and defined its offensive concept of operations based on the armor and air force, and the doctrine of static defense as anachronistic and counterproductive. As a result, significant professional debate about force buildup gathered momentum in the IDF between two groups, broadly defined as "traditionalists" and "reformers."[17] Traditionalists went along with the conventional wisdom. In future war, they argued, Arab armies equipped with vast amounts of modern weaponry would try to impose on Israel a war of attrition. To counterbalance Israeli "weakness" to wage a prolonged campaign, they proposed the reliance on a traditional swift offensive that would bring battlefield decision, and criticized defense as a negative form of battle. Preventive offensive was seen as preferable over even the shortest defense. Tanks coordinated with airpower, and assisted by other corps, were seen as an ideal means for maneuvering, attacking, and capturing and controlling territory. Driven by this line of thought, "traditionalists" saw in the huge armor formations the backbone of future force. They demanded a sophistication of the iron fists of the IDF that would bring the offensive deep into the enemy rear. They did not ignore new technologies; they saw in them promising force and protection multipliers against enemy countermeasures.[18]

The reformers offered an alternative for the offensive breakthrough battle, which they regarded with skepticism. Instead of breaking through an almost impenetrable Arab defense, they proposed exhausting enemy forces and inflicting heavy losses in the front and in the rear by using air force, navy, and special operations forces and by capitalizing on Israeli superiority in stand-off PGM, command and control systems, and target acquisition capabilities. They argued that by utilizing a qualitative edge in technology and human skills, and embarking on the defensive concept of operations, Israel could attain significant strategic benefits. Subsequent envelopment maneuvers against a weakened enemy would be far less impressive than the victory in 1967 but would minimize attrition rates.[19]

The debate between the two groups expanded beyond the IDF and rose to a national level. In 1987, Dan Meridor from the Foreign and Defense Affairs Committee of the Knesset recommended gradually decreasing the traditional armored force and injecting free resources into the massive procurement of stand-off PGMs. This, the committee concluded, would create qualitative asymmetry in favor of Israel. Minister of Defense Yitzhak Rabin was open to the proposals of the reformers; he was reluctant, however, to discredit the offensive military doctrine. Thus, seizing the middle ground between the two schools of thought, Rabin chose to develop and to procure new weaponry, but to incorporate it in the existing force structure and concept of operations, without effecting any changes whatsoever in the architecture of Israeli military power.[20]

Additional imperatives pushed toward further transformation, when the confluence of several political, social, and economic developments began to redefine the Israeli approach to military affairs in the early 1990s and led to a search for alternative solutions to preserving Israeli military power.[21] The peace process and the future return of the territories to Arab states potentially nullified Israeli operational depth and made the traditional instinct for preventive strike and taking the war to enemy soil problematic. Under these geostrategic changes, the IDF was compelled to look for doctrinal alternatives to preserve Israeli security.[22] In addition, Israel was incrementally transformed from a collectivist society to an individualist market economy with a higher priority placed on nonsecurity issues and a reluctance to spend excessively on defense. Under liberalization of the Israeli society, the idealistic zeal of both recruits and reservists to invest time and energy decreased dramatically, and voices for abandonment of universal military service were heard. Economically, massive procurement of high-tech weaponry became extremely expensive in terms of training and compensation of the force that operated them. In tandem with the above processes, lessons from American military conduct in the first Gulf War began to permeate the Israeli vision. Desert Storm offered, in the view of many Israeli experts, an example of a major advance in "sophisticated conventional warfare" and demonstrated how advanced technology may revolutionize war.[23]

In light of these social and geopolitical changes, and under the influence of the American experience, the Israeli military establishment started to contemplate the notion that big military and advanced technologies might be mutually exclusive, and that the best way to exploit the full potential of modern weaponry would be a small, sophisticated and professional force. In the process of digesting lessons from the 1991 Gulf campaign, the IDF gradually assimilated the idea that offensive mobility on its own was not an ultimate solution for the modern battlefield.[24] One after another, Israeli chiefs of the GS proclaimed their intent to transform the IDF into a "small and smart military."[25] Their ideas strongly echoed the arguments of the "reformers," who called for defensively oriented doctrine, and recreated a wave of attacks on the traditional security paradigm. The reformers argued in favor of destroying the enemy deep inside its territory without crossing international borders, and maneuvering precision fire in place of heavy forces.[26] Mostly from outside the IDF, experts examined the interrelation of stand-off PGMs, advanced sensors, target acquisition and C4I systems, UAVs, electronic and information warfare systems,[27] and their impact on the nature of warfare.[28] Under the rubric of "the future battlefield," they analyzed how operational and strategic visions could be reassessed under the impact of these new technologies when linked to a creative concept of operations.[29]

The overall influence of this conceptual debate about the changing nature of warfare on the defense establishment was incremental. However, as an institution, the IDF made only piecemeal attempts to comprehend the emerging security regime and to transform itself along those lines.[30] Because of the intellectual conservatism and lack of relevant organizational framework, the IDF was unable to put these concepts together in a coherent form and to translate the abstract theoretical ideas into concrete military reforms. It was integrating technological advances into existing organizations and routines, rather than experimenting with radically different concepts. The acceleration of the Israeli RMA depended on the personalities involved, no less than on the changes in the strategic environment. Not until the time when a group of reformers determined to implement these ideas institutionalized itself in the IDF was the Israeli RMA put into motion.[31] In contrast to "civilian" reformers outside the IDF, who mostly criticized the "offense and heavy" orientation of the IDF doctrine, "insiders" argued that lack of developed military thought and operational cognition produced this conceptual myopia. During the 1990s, three elements were introduced simultaneously into Israeli military praxis: a concept of operational art, a unique methodology for the analysis of military operations, and acknowledgment of the RMA. These three elements became organically interconnected: by utilizing new methodology, the IDF embarked on organizational and conceptual RMA innovation and treated it as development of the Israeli version of operational art. All three elements were introduced by the same small group of "insiders."

An Acknowledgment of the RMA: Crafting the Israeli Operational Art

Serious study of the RMA, and introduction of this concept into military praxis, got under way inside the IDF with the establishment of the Advanced Operational Group, later upgraded to the Operational Theory Research Institute (OTRI). In the early 1990s, several retired Israeli generals and academics in the IDF military college came to the conclusion that, occasional display of operational insight notwithstanding, the IDF totally lacked *operational thinking*—an essential domain of knowledge between tactics and strategy. This void, according to the group, resulted in the conservative approach of Israeli military thought and prevented adaptation of the IDF to emerging technological and geopolitical realities. Short of operational cognition, the IDF lacked the intellectual milieu for systematic thinking about organizational structure and concept of operations. The group resembled the above-mentioned "reformers" and criticized both the form (the lack of an intellectual approach) and the essence (the doctrine's irrelevancy) of the Israeli military praxis. They criticized the idealization of a tactical offensive excellence, as opposed to the development of a proper theory of war based on operational logic. According to them, tactical excellence in armor warfare diverted professional awareness from broader aspects of the theory behind operational maneuvering. They stressed the predominant tendency among IDF commanders to perceive the desired outcome of any combat activity as "mechanical destruction of opposing force." They criticized traditional armored offensive maneuvers, which could neither deter nor prevent the threat of a surface-to-surface attack on the Israeli rear or low-intensity conflict.[32]

The group sought to embark on scientific research to fill these voids, and its intention was approved by the GS in 1994. OTRI saw itself as a think-tank and consultancy for the development of an Israeli version of operational art; as an "experiment laboratory" for innovative concepts and doctrines; and, finally, as an "educational order" that would enlighten the IDF commanders to think critically and systematically about military affairs.[33] The institute became the

"knowledge authority" of the IDF in the decade that followed, and its influence grew tremendously, along with the improvement of its bureaucratic position within the military. The first educational activity for senior IDF officers,
known as the Advanced Operational Course (KOMEM), was approved by the
GS in 1995 and conducted the following year. By 1997 the Central and the
Southern Commands employed the conceptual tools offered by OTRI, and by
2000 the GS itself cultivated and disseminated the institute's ideas in the IDF.
The intellectual engines of OTRI were Brig. Gen. (ret.) Shimon Naveh, who led
the institute alternately with Brig. Gens. (ret.) Dov Tamari and Zvi Lanir. The
institute stood behind most of the organizational and conceptual transformation that gathered momentum within the IDF in the subsequent decade.[34] In
its activities OTRI advanced along two vectors: *operational art,* as a major research interest, and *general systems theory,* as a major methodology of inquiry
and conceptualization.[35] The methodological credo of OTRI included theories
from *architectural design* and *postmodernism.* The quintessence of these concepts served as the intellectual basis for the formulation of the IDF Concept of
Operations—the codified version of the Israeli variation on the RMA theme.[36]

Until the early 1990s, similar to the situation in the US prior to the 1980s,
the importance of the operational level was absent from professional cognition of the IDF.[37] Filling this void, OTRI was inspired by two theories of modern military thought: the Soviet theory of Deep Operations and the American
doctrine of the Air-Land Battle. Because the latter was seen by OTRI as an
American emulation and adaptation of the former, Soviet works on operational art became the yardstick of theoretical excellence[38] and the intellectual
fundament for Israeli experts.[39] They saw the Soviet terminology as much
more advanced and rich than any Western analogue,[40] and applied it extensively to interpret Israeli military affairs and doctrine.[41] After 2000 the documents and memoranda written in the GS, as well as the professional jargon of
Israeli senior officers, included a significant quantity of terms borrowed from
the Soviet military lexicon. OTRI experts saw the Soviet and the American
doctrines as methodologically identical: both recognized the existence of
operational art; both developed "harmonious patterns of thought and action
in an environment saturated with contradictions"; both had put away "a mech-
anistic approach aimed at the destruction of enemy's forces" and had chosen a
"systemic approach seeking to disrupt the rival system's operational rationale."[42]

OTRI proposed to emulate this approach for the development of Israeli operational doctrine. This methodology envisioned the enemy as a multidimen-

sional system, whose ability should be neutralized by: *fragmentation strike,* which isolated an enemy's military subsystem from the strategic supersystem and disrupted its consolidating logic; *simultaneity,* which coordinated actions across the spectrum of operations, to shock and paralyze an enemy system; and *momentum,* exploitation of the synergetic effect produced by fragmentation and simultaneity, in order to deny the opposing system response time, ultimately causing it to break down.[43]

Israeli experts argued that the Soviets and the Americans had succeeded in making this conceptual leap forward, thanks to adaptation of the *systemic thinking approach.* Conceptualization of warfare in general systems language provided, according to OTRI, the most relevant analytical lenses to deal with a complex and chaotic operational environment. They believed that this theory would enable translation of abstract strategic directives into mechanical tactical missions, and vice versa—linking all the tactical engagements together to achieve the strategic goal.[44] The battlefield decision, according to this school of thought, was not necessarily occupation of territory or destruction of enemy forces in an integral battle of annihilation, but neutralization of an enemy system's logic by the triple operational strike described above.[45] According to Israeli experts, this logic was applicable to both low- and high-intensity conflicts,[46] and the digitalization and extension of the battlefield in the RMA era only multiplied the relevance of this approach.[47] After the mid-1990s most of the Israeli senior officers went through KOMEM, an educational course offered by OTRI. The curriculum consisted of four major thematic blocks: *operational art,* based on the works of the Soviet theoreticians; *structure of paradigmatic changes, epistemology and dialectics; systems theory and cybernetics;* and *space perception.*[48] The graduates of their educational activities were seen by OTRI as soldier-philosophers who would become agents of influence inside the senior level of the IDF.[49]

OTRI experts presumed that the main asset of future military forces would be the ability to rapidly evolve knowledge, and they sought to provide the IDF commanders with unique, systematic reasoning capabilities. The methodology produced by OTRI was a mixture of general systems, chaos, and architecture theories, and postmodern approaches from various academic fields.[50] OTRI experts saw rationalism, scientism, and objectivity, where formal logic reduced the perception of the world to its elementary forms in order to pinpoint causality, as irrelevant mechanical reductionism. Like postmodernists, they maintained that reality is an organic, nonlinear, complex, self-organizing, dynamic system of interacting items, where chaos exists side by side with order. Warfare was seen as chaotic engagement between two systems and as an ephemeral accumulation of events, where each situation was different and unique, and open to a variety of interpretations. OTRI argued that military organizations, with their hierarchical, linear mind-set, tend to perceive complexity as a simple system, applying a "one doctrine fits all," mechanical approach. OTRI offered to grasp reality in its totality and internal dynamic, without reducing its complexity. To adapt military doctrine to the uncertain and changing reality, they legitimized *subjective knowledge in context* as the only tool to cope with complex issues. To achieve this goal, OTRI's methodology advocated *conceptualization*—developing an invented language to explain observed phenomena in the given context.[51] It did not accurately represent an observed phenomenon but made heuristic interpretations about it.[52]

OTRI saw the traditional IDF approach to "estimation of situation" as mechanical and irrelevant for operational and strategic planning.[53] Conceptualization, in its turn, would enable interpretation of blurred political directives, and their translation into effective operational campaign planning. Conceptualization was expected to enable "discourse"—a dialogue between the political echelon and the operational and tactical level of command, where everyone would be able to interpret intentions and orders of others.[54] The ideas generated during the discourse were presented in "knowledge maps," and the insights were translated into operational directives and orders.[55] OTRI trained the IDF commanders to see themselves as "operational architects"—to interpret and to conceptualize combat environments using the language of avant-garde architectural schools and works on urbanism, cybernetics, and psychology. It was OTRI's belief that the logic of these disciplines offered greater facility for interpreting situations than did traditional military terminology. Theoretical texts considered essential readings were works by Gilles Deleuze, Felix Guattari, Christopher Alexander, Gregory Bateson, Clifford Geertz, and Bernard Tschumi. The language used by senior Israeli commanders in this period resonated strongly with the architectural literature. For example, urban warfare in April 2002 was described by the Israeli commanders as "inverse geometry."[56]

OTRI's ideas spread across the IDF rapidly. In 1998, the head of the Central Command, Moshe Ya'alon, coping with challenges in the Palestinian arena, substituted "estimation of situation" with a new alternative methodology. The Central Command invented its own tools, language, and concepts to examine phenomena, trends, and occurrences beyond the tactical dimension of the conflict.[57] This new methodology moved to the Southern Command, then to the GS, and from there it was disseminated throughout the IDF, including the Research and Analysis Directorate of Military Intelligence. When Ya'alon became the chief of the GS, he turned the "conceptualization" and "discourse" process into "a main axis of the GS estimation of situation."[58] Ya'alon saw this methodology as a tool for generating a conceptual and organizational revolution in the IDF that would introduce new military knowledge into the intellectual vacuum on the operational level.[59]

When in the mid-1990s the American theoretical RMA debate had evolved into the all-encompassing military reform, the Israeli Ministry of Defense decided to cooperate on this issue with its US colleagues. ONA was identified as a main intellectual and bureaucratic engine of that process in the US, and designated as a counterpart. OTRI was chosen as the most relevant Israeli partner.[60] The first encounter between the RMA theoreticians from the two countries took place in 1997 in Israel. Its goal was to expose each side to the other's professional vision of the changing nature of war. American experts headed by Andrew Marshall presented the main postulates of the American approach to the RMA.[61] OTRI and the Doctrine and Training Unit of the GS (TOHAD) observed the changing security and technological environment, recognized the necessity of embarking on a profound organizational and conceptual reform, and presented a new methodology for the development of Israeli military thought and operational art.[62] The Israelis did not accept the term *RMA* in its pure American form, having found some of its aspects irrelevant to the operational realities of the IDF.[63] However, the RMA concept on the whole was found extremely relevant for emulation. Israelis who participated in the meeting saw themselves as the architects of the Israeli version of the RMA. Although differences of interpretation were in evidence, both delegations acknowledged that the emerging transformation in the nature of war should be reflected in the organization and concept of operation of the two militaries. It was agreed that in subsequent years the experts of the two countries would be in constant touch, for adaptive, mutual learning.[64] The subsequent contacts, workshops, and scientific seminars on military theory, innovation, and experimentation with the ONA[65] became one of the most important sources of intellectual influence on the builders of the Israeli RMA concept. The concept of RMA and, later, the US Defense Transformation were presented by OTRI as the main frame of reference for similar processes in the IDF.[66] Israeli conceptual reform, which resulted in a new Concept of Operations, was deeply influenced by the US Defense Transformation.

Following the first meeting with ONA, the Israeli participants started to cultivate an intellectual climate in the IDF that would foster acceptance of the RMA.[67] About a year prior to the first meeting with the Americans, when interest in the topic of the RMA became officially established at the IDF, the Doctrine and Training Division, influenced by OTRI, started to translate and to disseminate historical and theoretical works on the Soviet MTR, the American ALB, and the American RMA, which aimed to fulfill educational lacunae in the theoretical knowledge of IDF officers.[68] Simultaneously, an unprecedented number of professional publications on various conceptual aspects of operational art appeared,[69] including applications of smart munitions and information technology on the operational level.[70] Articles translated from English reflected the American vision of the RMA new theory of victory.[71] Following the first encounter with ONA, the opening volume of *Ma'arachot* in 1998 was fully dedicated to the RMA[72] and its implications for Israel.[73] This theoretical knowledge about diagnostics and changes in the nature of war was designated as an essential intellectual foundation for formulating the Israeli version of the operational art. In subsequent years manifestation of the American RMA during the conflict in Kosovo was vigorously discussed in the IDF. Technological and conceptual lessons from the American campaign significantly influenced the Israeli visions of modern war, especially in terms of smart munitions and the role of airpower in the modern battlefield.[74] It became obvious that technological, political, and social changes made the existing paradigm of military operations obsolete and that there was a need to invent a new system of explanations. However, in 1997, OTRI experts, paraphrasing Thomas Kuhn, argued that in terms of the RMA the IDF was in a stage of the epistemological crisis—one step before emergence of a new paradigm.[75]

The New Concept of Operations: Theory and Practice
In the late 1990s the IDF higher command had realized that the combat paradigm of the IDF had lost its relevance for the emerging assortment of scenarios and threats. The need for a new concept of operations arose from strategic changes in the region, diversification of the balance of threats, budget cuts, advanced technological capabilities, and the influence of the American RMA's ideas about the changes in the nature of warfare. Synthesizing between two sources of intellectual inspirations (Deep Battle of the Soviet operational art and the US ALB) and elaborating on the methodology of general systems theory, OTRI started to make radical proposals for organiza-

tional and conceptual changes in the IDF, beginning in the late 1990s. The most central recommendation demanded the formulation of an operational doctrine that would serve as a coherent framework for the buildup, training, and operation of the IDF. Specific recommendations included the reduction of traditional elements of heavy maneuvers and development of operational capabilities for mobility and striking power and for waging simultaneous multidimensional operations on the front and in the rear, both by fire and by maneuver.[76]

Heads of the regional commands, influenced by OTRI methodology, began posing questions regarding planning of military operations, organizational structures and force buildup, and command and control relations with the GS. In response, then-CGS Shaul Mofaz set up a team that drafted a document entitled *IDF Strategy*. However, the work got bogged down and was renewed under CGS Ya'alon when, in 2002, he established several teams of senior officers to reassess various aspects of the Israeli military doctrine. Ya'alon saw this process as a conceptual revolution in Israeli military affairs that would provide doctrinal guidance for the IDF in the RMA era. The new Concept of Operations (CONOP) published in 2006 became the quintessence of Israeli views on the nature of warfare in the RMA era.[77] Since OTRI experts provided professional-academic support for doctrinal workshops, the final product reflected the significant number of its ideas promoted, including the methodology of systemic operational design (SOD).[78]

The new CONOP prescribed transforming both the organizational structure and the doctrine of the IDF. In terms of fire–maneuver balance, the CONOP emphasized precision and stand-off firepower (mostly from the air) at the expense of ground maneuver by heavy armor formations. It discredited the utility of massive ground engagements and defined capturing territory as less important than destroying targets with minimum casualties and collateral damage. Holding territory was deemed a political and operational burden. Presence on enemy soil was to be replaced by the ability to maneuver through it for fixed periods of time, and to produce "operational effects" to impact an enemy system. In contrast to the traditional IDF doctrine, the CONOP advocated bringing the war to the enemy's territory not by "boots on the ground," but by an integrated system of stand-off, precision-guided fire, based on real-time intelligence and supported by command and control systems. These reconnaissance-strike complexes were expected to enable stand-off control of the territory.[79] The actual capabilities for these tasks existed for more than a decade;

however, military thinking had not kept pace with the sophistication of the IDF equipment.[80]

The new theory of victory strongly contrasted with classical linear operations, where the enemy was brought to its knees in a decisive battle of annihilation. The new approach demanded simultaneous attack throughout the entire depth and dimensions of operational deployment that would create overall paralysis of the enemy system. This type of victory not only was based on physical annihilation, but paid a great deal of attention to affecting the "rationale" of an enemy system and paralyzing its motivation and ability to keep on fighting.[81] In addition to emulating the Soviet approach, which is clearly seen here, this new theory of victory relied heavily upon the American concept of Effect-Based Operations (EBO). EBO advocated applying military levers not only for the sake of inflicting damage per se, but in order to produce indirect and cascading effects that would influence the enemy as a system and attain the strategic goals of the campaign. In 2003 OTRI organized a conference on EBO and, together with TOHAD, disseminated explanations of its operational principles in the IDF. The GS workshops identified EBO doctrine as a revolutionary military concept, offered training for it during the large-scale exercise in summer 2004, and adapted it in the new CONOP.[82] In its emulation of the US, the IDF was strongly influenced by the application of these new ways of war in Iraq.[83]

Although the new CONOP saw precision and stand-off fire as the principle technique of future warfare, maneuver was not totally discredited. The CONOP suggested establishing a new force structure for the IDF basic formation. It was envisioned as a network of small and autonomous units that would be capable of executing versatile missions, including target acquisition and guidance and execution of precision fire from the air and from the ground. These units would operate as a self-synchronizing network, in which formations coordinate with each other horizontally, and orient and navigate through the battlefield without going through the central command. A network of separate units would be diffused throughout the whole operational depth and would operate semi-autonomously, but jointly in frames of the unified campaign plan. Under the assumption that it takes a network to combat a network, these "swarming" formations would adjust themselves to the stealth capability of the enemy. Considerable authority would be delegated top-down, and decisions would be based on chance, contingency, and opportunity and would be made in real time, on the immediate tactical level. This concept was expected to be effective against both conventional and asymmetrical threats.[84] Although developed in-house, "swarming" and diffused warfare tactics were strongly inspired by the US Network-Centric Warfare concepts. Given the frequent contact with the parts of the US military responsible for the development of and experimentations with *transformation concepts,* it is not surprising that the IDF was strongly influenced by this American approach.[85]

To ensure that combat activities could be simultaneously executed by all the involved IDF units and synchronized throughout the whole operational depth and from all possible dimensions,[86] the concept of *jointness (shiluviut),* borrowed from the US Defense Transformation, was established and became another banner of the Israeli RMA. It implied the ability to deploy units from all of the branches in a synchronized manner, which became possible through the advanced command and control systems. Ideally, *jointness* blurred borders between services, left classical linear combat behind, and opted for simultaneous multidimensional warfare.[87] The new CONOP also added an operational level to IDF planning and thinking. The concept of an "operator"—a commander responsible for deployment of all the forces in the comprehensive theater of war—was introduced into the GS alongside the notion of *jointness.* New departments for campaign planning were established in addition to the existing operational planning units.[88] The preference for flexibility in weapons development, in light of diverse and unpredictable threats,[89] also strongly resembled the recently introduced American "capabilities-based approach."[90]

The operational experience during the Second Intifada became a laboratory for some of these new approaches.[91] The combat experience was regarded as successful and provided field-based evidence in support of the new CONOP and the methodology developed by OTRI.[92] Diffused warfare was considered a great success and an effective revolutionary innovation.[93] The IDF operated in the West Bank and Gaza Strip as a network of reconnaissance-strike complexes, which collated data on highly dynamic targets from an assortment of intelligence sensors and transmitted it in real time through sophisticated C4I systems, to precision-guided, stand-off, "smart" fire from manned and unmanned, aerial and ground platforms. The IDF shortened sensor-shooter loops to minutes and seconds. It widely practiced the jointness concept, when these sensor-shooter centers consisted of representatives from different IDF branches, and even civilian intelligence agencies, and operated as an integrated whole.[94] Eventually, the GS convinced itself of the rightness of its course. It was believed that these operational methods could be applied to any contingency scenarios, both in low-intensity conflicts (LICs) and in high-intensity conflicts (HICs).[95] Advanced systems, which would enable the application of these concepts to the conventional realm, were rapidly developed.[96]

According to the common wisdom that crystallized after the 2006 campaign, the ideas inspired by OTRI account for some operational misfortunes during the war, and much blame for the poor performance of the IDF lay with the recent CONOP.[97] This work deliberately refrains from analyzing the combat effectiveness of the CONOP in the context of the Lebanon war, primarily because of the different research focus. The book deals with the intellectual history and cultural roots of military innovation, regardless of its combat utility. However, some remarks are appropriate here. When judged from a more distant perspective, it seems that linking the poor combat effectiveness of the IDF directly with the 2006 CONOP and OTRI is an oversimplification. The Lebanon campaign cannot serve as an empirical test for the quality of the IDF's "operational theory," because it was not really implemented in Lebanon. The CONOP was the latest, but not the final, stage in a decade-long process of crafting Israeli operational art. Moreover, the document was introduced to the IDF several months prior to the war, and its concepts still had not gone through field experimentation and a full training circle when the war broke out. The concepts were disseminated through the IDF but not fully absorbed; they were vaguely understood and implemented only partially during the war, and not according to doctrinal guidance.[98] Also, it should be emphasized, OTRI developed theories and concepts that were intended to serve as thinking tools for the IDF and not as command practice.[99]

Indeed, the critique of "operational theory" both before and after the 2006 campaign and the poor conduct in the field reflect a conceptual cacophony that emerged in the IDF prior to the war. Many critics outside the IDF treated the newly invented language and methodological approach as esoteric, blurred, incomprehensible, and meaningless. The uncontrolled development of the new terminology, free from any standardization, interfered with the ability to grasp the essence of the operational concept. Even if potentially contributing, innovative methodology and doctrinal ideas were lost in lexical confusion.[100] The euphoria of conceptualization "befuddled the minds of Israeli military leaders"[101] and turned the IDF into a Tower of Babel, in which tactical commanders had no clue during the war about how to translate incomprehensible orders formulated by their superiors in postmodern language.[102] Other experts questioned the very essence of the innovation. Some defined the doctrinal logic of the new CONOP as counterproductive, compared to classical ways of waging war,[103] and saw the application of postmodern concepts to military affairs as damaging, if not downright unprofessional.[104] An additional group criticized the blind, mechanical emulation of the American RMA, which was seen by them as irrelevant for the Israeli geostrategic realities.[105] Some senior officers argued that there was nothing revolutionary in the changing military regimes, with no impact whatsoever on the classical principles of war.[106]

During the several years prior to the war, a significant number of senior IDF officers approached new concepts from the position of bureaucratic and intellectual conformism and utilized the new vocabulary only as lip service to the conceptual fashion of the moment. Very few truly studied these new ideas deeply and in an orderly fashion. Even fewer were ready to admit that they did not understand the meaning of what they were reading and/or saying. While some voiced their concerns, others were embarrassed to openly admit their confusion. Using the same terminology they often meant different things, which contributed to the overall conceptual mess. In 2005 the new CGS, Dan Halutz, started to reorient the IDF in a somewhat different conceptual direction. With his arrival, the IDF was further disoriented conceptually and tried to figure out which of the concepts associated with the previous CGS's school of thought stayed intact under the new boss and which should be thrown away. The confusion reached its height when in April 2006 CGS Halutz, who declared his intention to redirect the IDF conceptually, approved the last CONOP, which he inherited from his predecessor without making any changes. The CONOP was defined as a basic military document, but it was not entirely clear within the IDF whether the document should indeed be implemented to the letter or should be treated as an additional new draft, like those disseminated in the previous years. When Israel surprised itself with the decision to go to war, the IDF was experiencing the climax of its conceptual disorientation. Eventually, different command authorities in the IDF approached warfare from somewhat different conceptual backgrounds, according to how they understood the notion of "operational theory" and its doctrinal status in the IDF. This bureaucratic-conceptual chaos and doctrinal relativism were among the factors that made the IDF war machine somewhat dysfunctional during the war.[107]

PART 2: ISRAELI STRATEGIC CULTURE

Cultural Characteristics and Cognitive Style

Although Israeli culture is a fusion of Jewish, European, subethnic Levantine, and American traditions, it is not accurate to describe it as a "shapeless hybrid about which all generalizations are impossible." Its coherent inner logic allows one to speak of a single Israeli culture.[108] In terms of social structure, individualism coexists in Israel with well-developed group orientation,[109] but that differs from "individualistic America" or "collectivistic Russia." Israeli individualism expresses itself in a casual attitude toward rules and regulations, in self-reliance, and in little respect for imposed authority. However, Israeli loyalty is less oriented to oneself and tends to be focused more on the pursuit of collective goals and the well-being of the society. Despite the ongoing transformation of the Israeli society from a collectivistic ideal to a more liberal individualistic one,[110] and Americanization of Israeli culture,[111] in Hofstede's measurements Israel scored only moderately individualist.[112]

Egalitarian social norms established by the founding fathers produced a striking informality of behavior and inattention to hierarchy. Defined by anthropologists as a "small power distance society," Israel has a second-lowest inequality in superior–subordinate relations.[113] This atmosphere exists even in such a hierarchical organization as the IDF. Having a lean organizational structure and simple military bureaucracy, the Israeli military and defense systems are informal and egalitarian and foster a lot of innovative ideas bottom-up through informal organizational shortcuts.[114] Lack of distance, a cult of simplicity, and social informality have produced an "ultra-low-context" communications style among native-born "new Israelis" (*sabras*). In *sabra* narrative a direct, even confrontational, communication style (speaking "*dugri*") was equated with integrity and synonymous with "honest and authentic." In contrast, a pacifying, diplomatic style is regarded as suspect and hypocritical and perceived as "insincere and artificial."[115] Although the various waves of immigration brought different cultures and communication styles, they were socialized to the norms of speaking *dugri* and to the point (*tachlis*)[116] and were eventually absorbed by the low-context approach of the Israeli melting pot.[117] Israelis prefer a direct, spontaneous, natural, and unrestrained speech that leaves little to interpretation.[118] For practical Israelis, style matters less than content; they are uncomfortable with formality and ceremony and disrespect external forms of courtesy. They have manifested this mode of interaction during negotiations with Egyptians, Syrians, and Palestinians.[119] As to the monochronic–polychronic dimension, Israelis tend to tolerate several things happening at once. However, they seem to be more multifunctional in deeds than polychronic in thought.[120]

The extreme dynamism of Israel's population has posed difficulties to formulating a solid diagnosis of its cognitive style. Cognitive psychologists have done relatively very little analysis of the Israeli culture. The existing findings suggest that Jews from oriental countries tend to be more field-dependent (holistic-dialectical) than their counterparts from the West (who are more prone to logical-analytical style). Consequently, scholars report on significant manifestations of both field-dependence and field-independence among the Israelis.[121] There is a solid basis to assume, however, that practical and pragmatic Israelis are more at home with logical-analytical approaches to processing data than with a holistic-dialectical grasp of reality, which favors descriptive knowledge. The orientation toward doing, coupled with the trait of anti-intellectualism, predisposed Israelis to value an inductive approach, in which praxis and procedural knowledge precede deep visions. As in the Torah statement *na'ase ve nishma* (we will do first and then we will hear), for Israelis in general and in military affairs in particular, doing comes before understanding.[122]

Israeli decision making is pragmatic, and military experts tend to view issues "from an analytical problem-solving perspective."[123] Given their pragmatic instrumentalism,[124] the acute sense of practicality, assertiveness, and tendency to improvise,[125] underscored by a low-context orientation, it is not surprising that mainstream Israeli thinking favors procedural over descriptive knowledge.[126] Pragmatic Israelis tend to focus on the essentials and to simplify everything, from organizations to procedures. Discussing the cognitive style of Israeli military professionals, Cohen, Eisenstadt, and Bacevich maintained that "Israeli operational concepts resemble keys carefully crafted to fit particular locks, rather than a general approach to the problem of opening doors." For this reason, "it is highly unlikely that there could be an Israeli equivalent of the *US FM 100–5 (Operations)*—a generic manual prescribing common principles and basic methods."[127]

Israeli Style in National Security and Its Culture of War

Although Israel has been embroiled in warfare more than any other country since the end of WWII, its approach to military strategy has been described as a "conceptual salad," marked by a lack of vision.[128] The IDF never formulated its strategic insights and military intuition into a comprehensive official written document until 2006.[129] Nevertheless, a set of widely shared national security stratagems drove Israeli strategic behavior.

Over the course of history, the Israeli approach to national security rested on three fundamental pillars: *strategic deterrence, early warning,* and *battlefield decision.* According to a basic assumption, the glaring imbalance of resources vis-à-vis the Arab states prevented Israel from gaining a terminal military victory. While Arab countries could afford to lose wars, a major military defeat would mean annihilation for Israel. *Strategic deterrence* was devised to prevent this scenario. If this first pillar collapsed, then, the Israeli intelligence community was expected to alert the political leadership to an impending conflict. The second pillar—timely and precise *early warning*—provided the necessary time to deploy the pool of reservists, while a small standing force was expected to block the enemy's initial advance. At this stage of Israeli decision making, consideration of preventive strikes came into play.[130] Finally, Israel was expected to materialize the most crucial of its pillars—*battlefield decision.* Although not intended as a terminal knockout, it was to bring the ultimate victory. The lack of operational depth created a conditioned reflex to transfer warfare to enemy territory. Casualty aversion and socioeconomic exhaustion demanded the attainment of victory as quickly as possible. A narrow political window of opportunity strengthened this Blitzkrieg reflex.[131]

The preference for short wars on enemy soil had a profound impact on the Israeli culture of war. Defensive, status quo strategy and offensive tactics became the regulatory norms of the Israeli strategic culture.[132] According to the shared belief, only an aggressive maneuverable warfare aimed to penetrate swiftly into the adversary's rear would make it possible to obtain a decisive victory in a short time.[133] Offensive operational initiative was believed to generate maximum efficiency from its smaller but better-trained, better-equipped, and more-motivated forces. Offensive tactics optimally exploited the synergetic effect of close air support, heavy armor, and intelligence and crystallized during the first two decades into an unwritten operational doctrine. The Wehrmacht's Blitzkrieg that served as the intellectual inspiration became an instinctive choice and an almost exclusive way of war. The leading professional periodical of the IDF published works on Blitzkrieg, despite the emotional difficulties involved in doing so.[134] The neglect of defensive operations in favor of offensive romanticism neutralized the notion of mobile defense in Israeli operational thinking, produced stratagems such as "static defense on stop lines," and diminished the ability to wage prolonged wars of attrition.[135] The ethos of *offensive à outrance* imposed a simplistic and mechanical approach on Israeli military thought. Dynamic during the first decade, since the late 1980s it became "an archaic dogma detached from both the political-strategic reality and the universal logic of the military profession." The basic assumptions were never reexamined, neither after the 1967 war when Israel acquired operational depth, defensible borders, and technological sophistication, nor after the 1979 peace treaty with Egypt, which reinforced these trends.[136] This cult of the offensive was in harmonic accord with such traits of the Israeli strategic personality as siege mentality, dynamism, initiative, improvisation, and operational creativity.

The Qualitative Edge and the Role of Technology

The acquisition of weapons more sophisticated than those of the enemy, and handling them more professionally, was called in Israel the "qualitative edge" and became a means to counterbalance the numerical superiority of adversaries. The quest for and preservation of this edge became the fundamental strategic motif.[137] In theory, the qualitative edge rested on the harmonic synthesis of material and morale factors.[138] Beyond technology, Israel traditionally emphasized high personal and educational qualities of conscripts, high motivation and unit cohesion,[139] and a creative operational stratagem,[140] sometimes referred to as an Israeli "indirect approach."[141] These three "soft multipliers," backed by state-of-the-art technology, became central to the IDF's theory of victory. In practice, however, Israel frequently saw in advanced technology a "quick fix" minimizing war costs, duration,[142] and fatalities.[143]

The role of technology in the Israeli culture of war grew steadily. The experience of being unable to obtain essential weapons in the face of isolation left a lasting mark on the Israeli psyche.[144] As Ben-Zvi has demonstrated, neither was the history of Israel's defense cooperation with its major ally one of total harmony.[145] Self-reliance in weapons production became a national obsession[146] and resulted in the establishment of a highly innovative defense industry.[147] In its early years, the IDF had little understanding of military technology and deployed obsolete, mostly pre–World War II, weaponry. In 1948, quantitative inferiority was compensated for by a higher caliber of Israeli soldier, creative planning, offensive initiative, and operational ingenuity of an indirect approach. During the 1950s weapons procurement improved, but lacked synchronization, was sporadic, and was poorly balanced across the services. Toward the

Six-Day War, Israel started to procure advanced foreign weaponry and to develop its own defense industry. By the 1970s it was totally self-reliant it terms of light weapons production. During the War of Attrition, state-of-the-art modern weaponry was used extensively. Even more superior technology was exploited in 1973. During the war in Lebanon, the IDF internally developed ultrasophisticated weaponry. Driven by a fundamental sense of insecurity and motivated by its technological success, Israel strove to attain full autonomy in weapons production regardless of the price.[148]

The Israeli approach to technology was divided from the beginning. On the one hand, the IDF always sought to be one step ahead technologically. On the other hand, this tendency was partially offset by the value placed on the human factor.[149] However, as van Creveld has pointed out, the emphasis on the stratagem began to vanish after 1967.[150] Following the 1967 war, the IDF regarded technological solutions more and more as a panacea in security affairs. Thus, sophisticated weapons systems began to dictate Israeli strategy. The more sophisticated the arsenal became, the more frequently this trait manifested itself.[151] During the 1950s "the IDF sought to achieve technological superiority where possible"; after the late 1960s the quest for technological dominance started to play "an important role in the IDF self-understanding," and by the 1980s the qualitative edge vis-à-vis the Arab world was measured mechanically in strong technological terms at the expense of other dimensions.[152] The instinctive Israeli preference for employing military means while dealing with national security problems[153] was increasingly accompanied, according to Bar-Joseph, by a conditioned reflex of looking for solutions through the prism of technology.[154]

Vigorous technological innovations coexisted in the IDF with a conservative military paradigm. Technical ambitions and interests of the defense industry frequently prevailed, in the vacuum created by the lack of an overall concept of operations.[155] While adept at tactical-technological innovations, the IDF was less capable of reconceptualizing its doctrine. Most of the IDF responses to its military failures took the form of quantitative-technological upgrades rather than qualitative-conceptual modifications. Israelis believed that mechanical improvements "in range, accuracy, or maneuverability could yield large differentials" in combat effectiveness.[156] Quantity gradually became a form of quality[157] and predisposed the subjects of Israeli strategic culture to believe that "what doesn't work with force will work with more force"[158] of technology. The traumatic experience of the 1973 war predisposed the IDF to believe that quality could not offset quantity by a force of its own and caused the IDF to almost double its order of battle.[159] As a side effect of this tendency, the necessity to pose fundamental theoretical questions about the art of war gradually diminished. This change in interpretation of qualitative edge in favor of growing reliance on technology encouraged the neglect of many nonmaterial dimensions and resulted in incremental debilitation of Israeli military thought.[160]

Siege Mentality and Aversion to Planning

A societal sense of insecurity, sometimes referred to as a "siege mentality," is another fundamental motif of Israeli strategic culture. For two millennia life in the Diaspora was an uninterrupted struggle for survival that became a central trait of the Jewish ethos. Enslavement, persecution, oppression, and systematic annihilation became pillars of the Jewish psychological milieu and have made a profound impact on the Israeli approach to national security. A siege mentality, a concern for security, and an obsession with survival underlay the Israeli strategic psyche, resulting in the operational assumption that the state of Israel is under constant existential threat.[161]

Ben-Dor emphasizes that collective memory of Diaspora shaped the Israeli siege psyche even prior to the traumatic experience with the neighboring Arab states.[162] The Arab–Israeli conflict was treated as a reincarnation of the traditional anti-Semitism.[163] Encirclement of Israel by enemies with greater populations, resources, and militaries further contributed to this strategic mentality.[164] Even after several generations had been born in the independent Israel, the ghetto mind-set of total insecurity and distrustful self-reliance was preserved as a fundamental narrative. The strategic mentality grasped international politics as a purely Hobbesian self-help system and made the quest for security stronger than the quest for peace.[165] Reflected in Israeli popular and political culture,[166] enhancement of security became a central social value, and struggle for national survival became the central narrative.[167] The collective sense of strategic fatalism[168] predisposed the defense community to emphasize worst-case scenarios,[169] and to seek absolute security,[170] and made a mentality of "yield no inch" a kernel of the Israeli strategic culture.[171] Although not a "garrison state," the security sector in Israel became primus inter pares, and the IDF was seen as the only guarantor of continuous existence.[172] Lack of achievement on the diplomatic front marginalized foreign policy and created a habitual preference for military levers over diplomacy.[173]

Valid or not, the self-image of Israel became one of "few against many."[174] In this atmosphere of constant siege, instability, and uncertainty, Israeli decision making tended to concentrate on short-run, day-to-day operational matters at the expense of comprehensive long-term strategic thinking.[175] The fear of perceived and actual threats resulted in providing ad hoc solutions to immediate problems.[176] Although this pragmatic approach had an advantage of flexibility, it dissuaded formulation of a "coherent and cohesive security doctrine"[177] and systematic staff work. Consequently, "tacticization of strategy"—replacing strategy with swift military improvisations driven by tactical thinking—became a common symptom of Israeli strategic culture.[178] According to Freilich, those experts who were not predisposed to immediate action and who tended to emphasize strategic complexities were marginalized by the mainstream Israeli strategic instinct.[179] Siege mentality bounded the horizons of Israeli military thinking and framed the psychological state of mind of Israeli military theoreticians. The IDF mostly focused on immediate operational missions of the current security (*batash*), at the expense of developing fundamental visions about the nature of war. That preoccupation with routine and ever-pressing missions left military leadership very little time to think about the future of warfare, or to experiment with new concepts. It discouraged the IDF from posing RMA-type questions on paradigmatic shifts in doctrine, organization, and concept of operations.[180] Rather than envisioning future wars, Israeli military experts were fighting the current ones.

Driven by the same ad hoc mentality, the IDF never codified security concepts and military doctrine in written form, under the assumption that political leadership would provide clear guidance for crafting the most relevant strategy in any given situation. However, Israeli leaders sought partial solutions for the strategic issues and were disinclined to formulate comprehensive strategic aims, driven by a "wait and see" approach. An additional reason for this short-term orientation was the unwillingness to announce long-term visions, which might sound painful and intolerable internally and decrease space for political maneuvering. Eventually, Israeli national security decision making managed grand strategy on a day-to-day basis.[181] The Israeli national comptroller issues reports almost annually exposing aversion to strategic planning, lack of long-term vision, and the preference for piecemeal, particularistic considerations.[182]

Culture of Improvisation

The Israelis have been characterized by cultural anthropologists, and see themselves, as assertive, flexible about plans, casual about rules, and improvisatory.[183] The latter refers to a characteristic of management, when a spontaneous recombination of knowledge results in problem solving, grounded in the realities of the moment.[184] In his works, Freilich shows how Israelis equate improvisation (*iltur*) with creativity, and proudly present a national capacity to do so as a hallmark of excellence in security affairs. According to Israeli societal beliefs, this approach guaranteed survival in an uncertain world and assured flexibility in times of crisis, without formulating objectives in advance. Exploiting the ability to orient, to think, and to bounce ideas around quickly, Israelis do not hesitate to find better, "not by the book" solutions, even if it means throwing out the existing operational procedures.[185]

The Israeli military tradition glorified the ability of its officers to quickly orientate, rely on personal judgment, think on their feet when confronting uncertainty, seize the initiative, and give on-the-spot solutions. Not allowing themselves to be controlled by the plan, flexibility, and improvisation are essential aspects of Israeli officers' training and became a more important virtue than acting according to prior planning.[186] Driven by this feeling of "national self-efficacy," military and defense officials deliberately sacrificed integrated planning for pragmatic improvisation. The latter enabled quick organizational learning, techno-tactical adaptation, and effective exploitation of short lines of communication.[187] The downside of this approach was amateurishness. The operational code of the Israeli defense community does not emphasize true strategic planning, nor does there exist the organizational culture or institutional capacity to do so. Detailed planning and staff work exist on the tactical-technological level of some services and in the defense industries. However, these islands of techno-tactical excellence are surrounded by the waters of strategic-operational mediocrity.[188]

The tradition to extemporize goes back to the early Israeli experience, when the young country was forced to deal with formidable security problems. Since the times of Ben-Gurion, strategic decisions were often generated in small ad hoc forums of a "kitchen cabinet" (*mitbachon*) in the discrete, oral, and not preplanned process. This uninstitutionalized and informal decision making enabled high-speed communication and problem solving.[189] The security situation before the War of Independence was so unpredictable that improvisation and crisis management came to compensate for long-term planning and deficiencies in numbers, weaponry, and training. Since the ability to improvise out of uncertainty proved successful, it was preserved despite the subsequent establishment of formal institutions and regulations.[190] During the Sinai campaign Israel functioned spontaneously on tactical, operational, and strategic levels and contradicted self-imposed procedures and regulations. The same improvisational opportunism manifested itself at the outbreak of the Six-Day War. The level of improvisation was significant in 1973, when the IDF prepared for one kind of war but fought another one, and both contradicted its (unwritten) military doctrine.[191]

The IDF leadership, too, has been characterized by "adventurous officers, demonstrating improvisation, initiative and flexibility, rather than going by the book."[192] According to the Israeli normative approach, the superior only gives general directions, leaving the commanders on the ground to translate the spirit of the command into action. Subordinates are encouraged to challenge ideas, suggest modifications, and come up with innovative solutions to problems. Otherwise, they might be perceived as lacking initiative, self-confidence, and backbone.[193] Despite a high level of techno-tactical achievement, the tendency to delegate authority and to reward initiative has led to accusations of dilettantism among the IDF commanders. Dynamism and improvisation frequently compensated for a deficit of professionalism on the operational-strategic levels.[194] Since then, the IDF has not divorced itself from what van Creveld defined as the Palmach culture of an "organized mess" (*balagan meurgan*),[195] where operational improvisation bordered on "free interpretations of mission objectives," and initiative bordered on operational disobedience.[196] It maintained a romantic, informal guerilla "count on me" (*smoch*) culture that rested on the belief that the clever ability to adjust to circumstances would get it out of any unforeseen trouble.[197]

This culture became one of the multipliers of the anti-intellectual approach in the IDF, because those who believed that they could deal with any problem on the spot had no incentive to learn lessons.[198] The IDF perpetuated this culture by promoting "improvisers" and "problem solvers," at the expense of military intellectuals.[199] Creativity and original improvisations coexisted in the IDF with self-satisfaction and impatience with the ideas of others. Reliance on the past was not considered a source of intellectual strength, but an out-of-context imitation.[200] In the realm of military thought the IDF has produced a culture of conservative and incremental innovation that has "resisted fundamental transformations."[201] The most damaging consequence of this tendency was disinclination and, eventually, fundamental inability to adapt conceptually in advance of changing military regimes.[202]

The (Lack of) Intellectual Tradition in the IDF

The Zionist vision of the "new Jew" strongly influenced the Israeli approach to military thought. For generations, Jews living in the Diaspora were excluded from agricultural and physical labor, and the bulk of the Jewish population were religious scholars. Religious passivity in the face of persecutions, and a victim mentality, prevented Jewish people from taking their fate in their hands. Zionist narrative opposed this reality and sought to create a new kind of personality that would be an antithesis to the Diaspora Jew—fearful, nonconfident, and overly conscious of authority.[203] The image of the "new Jew" (*sabra*) was of an "elect son of the chosen people," powerful and free of inferiority complexes, who would build his own life on his own land. Zionists sought to establish a large and productive working class, with only a small group of religious and secular intelligentsia dependent on it, in contrast to most of Diaspora history. Pursuing theoretical knowledge was rejected as inappropriate social behavior associated with weakness and unproductiveness. The European glorification of members of the intelligentsia was jettisoned in favor of the simple, physically strong, and straightforward farmer, who, like the Russian *muzhyk,* feels attached to the land he works and is ready to defend it. The "People of the Book" became the "People of the Plow and Rifle."[204] Israelis saw themselves as a nation with guts, which does not play by the rules, tries the impossible, and has a relatively high tolerance for risk taking.[205] Impudence (*chutzpah*), in the Israeli interpretation, acquired a positive meaning of nonconformist and gutsy audacity.[206] Israeli strategic culture celebrates the spirit of "can-do" leaders who push forward, disregarding the hardest obstacles.[207]

Israelis cultivated a self-image of "doers," rather than "talkers," not of philosophers but of pragmatic and realistic men of action who prefer deeds over words.[208] Pursuing higher education versus building the state of Israel was presented as one of the moral dilemmas of the early settlers, with a normative social preference for the latter. One of the by-products of this norm was a tolerance toward an anti-intellectualism of successful practitioners, which was cultivated and institutionalized.[209] In keeping with a general rejection of the "miserable bookish scholars" in favor of the heroic, fighting farmer, Israeli military culture preferred practitioners over theoreticians. Both groups existed in the IDF from the start; "practitioners" were primarily made up of members of the Palmach (the striking companies of the Jewish military force during the British mandate). The "theoreticians" consisted of former officers of European and Soviet militaries.[210] The Palmach self-consciously adopted a different, partisan-like culture that contrasted with the professional military. Although dilettantes in the professional sense, Palmachniks were vigorous tactical autodidacts, learning from their own mistakes and adapting elements from the experience of others.[211]

The founding fathers managed to squeeze maximum military effectiveness from the combination of creative and energetic *sabra* practitioners and foreign-trained military intellectuals. When the time came to establish a regular army, the latter were called in to formulate operational and organizational concepts, which with minor changes would serve the IDF for the next fifty years.[212] The balance between the two groups was, for the most part, preserved until the early 1960s, when the "performance-oriented" heroic *sabras* began to outnumber the disappearing generation of the intellectual veterans of the foreign militaries.[213] The subsequent decline of early theoretical activity and military thought within the IDF was in response to this generations change.[214] The "small war mind-set" of the guerilla-type raids of the 1950s predisposed professional norms and values of the IDF to favor "practitioners." The "raiding unit's style of command" glorified tactical activism and developed the ethos of aversion to thorough planning of operational rationales. An organizational climate formulated then would condition the operational and intellectual conduct of the IDF for the following decades.[215]

In her path-breaking research of Israeli military culture, Vardi argues that the atmosphere of intellectual stagnation and the lack of an "art of war" have been recurring phenomena in the IDF almost since its establishment.[216] Kober convincingly shows that the golden age of Israeli military thought barely extended beyond the first decade after independence.[217] Although initially the IDF had an operational-tactical concept that relied on the wise Ben-Gurion's geostrategic assessment, the IDF never really developed excellence in military thought. The primary designers of the Israeli approach to security and military affairs derived inspiration from the unorthodox but guerilla-type thinking. Educated on this ethos, and inspired by early successes, the IDF officers skeptically mistrusted grand theories, citing Clausewitz on the inevitability of a fog of war.[218] The tactical military achievements gained in 1949, 1956, and 1967 "created a self-image of continuous success." Following 1967 the leading characteristic of the Israeli generalship was a declared anti-intellectualism and a total disinterest in the art of war.[219] After the Six-Day War, the heroic practitioners' ethos was publicly and professionally glorified. Case studies of combat success served as the primary source of military wisdom and became a guideline for future military campaigns. The early intellectual culture was left behind. The neglect of abstract theory developed from a de facto situation to official policy.[220] Subsequent generations of commanders highly valued courage, persistence, and combat achievements, seeing leadership and not theory learning as an essence of military art.[221] Heroism, bravery, and the rich combat experience rather than theoretical knowledge or formal education became the recipe for successful problem solving.[222]

The arrogance that prevailed after 1967 further decreased the ability for abstract theoretical thinking.[223] Nurturing innovative, heroic, and improvisational instincts, the IDF became at best conservative, at worst unproductive, in the conceptual field, neglecting the intellectual aspects of military tradition.[224] Kober, in his study of continuity and change in Israeli military thought, argues that most of the IDF's intellectual vigor concentrated on the techno-tactical-logistical issues that give solutions to specific problems, at the expense of a comprehensive inquiry into the nature of modern war. This techno-tactical focus accounts for the poor performance[225] and a very late acknowledgment of the operational dimension of war.[226] The IDF lacked operational thinking—a level of abstract conceptualization that cannot be acquired through combat experience alone. Motivation, improvisation, and tactical assertiveness do not facilitate thinking in operational terms, which demands formal and continuous military education.[227]

In contrast to the eclectic founding fathers of the IDF, the heroic and charismatic Israeli generals exhibited little intellectual curiosity in studying military science.[228] Promotion to high-ranking positions notwithstanding, they never made the switch to strategic thinking and applied the same tactical logic when dealing with strategic issues.[229] Because the IDF has placed practice before learning and has been weakly inclined to study modern war in a multidimensional manner, a great gap had originated between theoretical and practical matters. Most of the IDF commanders rarely see military reading or theoretical writing as important.[230] According to Kober, in its approach to warfare the IDF preferred "to think in terms of technological force multipliers, such as smart weapons, than to create force multipliers based on smart doctrines."[231] It has been less innovative in its capacity to develop new fighting concepts than in its capacity to "capitalize on the relative inability of Arab militaries to improvise amidst confusion on the battlefield."[232]

Another reason for anti-intellectualism stemmed from the antiprofessional approach to military institutions. From the initial stages, the Zionist leaders sought to establish an effective but unprofessional army, strong enough to protect the Jewish people, but not so strong as to turn Israel into a garrison state. This view was based on a traditional Jewish philosophy that saw war as a choice of last resort.[233] Despite the populist military activism, Israeli leaders preserved the Jewish legacy of nonbelligerency by establishing the "people's army," not only as an instrument of security, but also as the national melting pot, responsible for various nonmilitary missions.[234] Strong critics of militarism, the early Zionists struggled to build a militia of "assertive warriors," but not a Prussian garrison state of "professional soldiers."[235] The normative self-view that rejected military service as an occupation resulted in the creation of an antiprofessional and subsequently anti-academic ethos of the IDF. This anti-institutionalism symbiotically merged with the military anti-intellectualism, which assumed that combat experience and charismatic qualities of the commanders could be a substitute for formal military education.[236]

New theories of victory usually emerge in "bookish militaries." Military organizations focused on material and technological aspects of war frequently neglect other dimensions of military theory that demand a greater degree of abstract thinking.[237] The existence of a group of educated officers "willing to think in a broad and disciplined way" about the nature of warfare, and capable of "making imaginative leaps into the future," seems to be a necessary condition for initiating paradigmatic change.[238] For most of its history, Israeli strategic culture lacked this phenomenon. The tactical smartness of the IDF was never translated into the intellectual inclination to make qualitative leaps in military thought or into the ability to craft a revolutionary theory of victory.[239] Israeli military thought lacked the ability to analyze, envision, and assess future developments and to generate knowledge in that context.[240] Although Ben-Gurion continuously warned that the most dangerous enemy of Israel's security might well be the mental inertia of its military,[241] the IDF failed to produce the correct equilibrium between muscles and brains.[242]

Promotion, Education, and Lessons Learned

Academic perspective on the military art, interest in systematic study of military history, and strategic thought never were part of the Israeli military ethos.[243] Intellectual brilliance in military theory never became a condition for advancement in the IDF, which promoted its officers for excellent performance on the battlefield.[244] Combat experience was the main parameter for professional evaluation; charismatic personality and native intelligence were regarded as more important than formal military education.[245] Most of the senior IDF officers did not undergo the training required for their posts, lack systemic professional education in the field of military science, and prior to their promotion have spent almost no time in professional military schools.[246] The tendency is to prefer an additional prestigious position in the field, over "wasting time" exploring the theoretical aspects of military art. The shared assumption has been that commanders will acquire all knowledge necessary for their careers through their experience—the more versatile the better.[247] Senior commanding posts were seen as positions that, unlike any other profession, did not demand formal academic education. The successful campaigns of 1948, 1956, and 1967 made this vision even stronger, since they proved that assertiveness, improvisation, and previous military experience can provide a solution to almost any battlefield problem.[248] Officers with an impressive combat record were promoted to key strategic posts, where, in keeping with what they had learned and assimilated, they performed in a techno-tactical manner.[249]

The IDF commanders' lack of military education and preference for personal experience over theoretical knowledge resulted in what was essentially the theoretical illiteracy of the IDF.[250] Israeli commanders assumed that professional intuition was preferable to studying military sciences, and had diverged from the conventional wisdom, which holds that practice depends upon knowledge, and not vice versa. The interest in military history and thought that originated in the IDF during the first decade gradually declined.[251] Even today the various IDF officers' training schools, biased by anti-intellectualism, are committed to the notion that "there exists no substitute for practical hands-on experience."[252] A university education, let alone a degree in military sciences, was not a prerequisite for officership or generalship.[253] Command and Staff College (*Pum*) never earned a reputation as an organization able to provide solid military education, à la the classical military academy.[254] National Security College (*Mabal*) served more for interorganizational integration than for thorough studies of military theory. Consequently, when Israeli officers arrive at positions where they should envision qualitative leaps in the nature of warfare, they lack basic intellectual training and theoretical skills.[255] The IDF, as organization, values university education of its officers but downplays professional military training in favor of practical civilian degrees—law, economics, management, and engineering. Lacking the institutionalized system of doctrinal training for its cadres, the IDF elite did not possess basic military theoretical knowledge.[256]

Although Israeli military culture has manifested a historically "impressive learning curve" in the techno-tactical realm,[257] the IDF did not develop any formalized systems for learning lessons from its campaigns. In most of its wars, it tended to repeat operational mistakes, and improvisational efforts usually reinvented the wheel.[258] Lessons were learned in a fragmented manner, not as a form of integrative analysis based on levels of war (strategic, operational, tactical), but parochially, across the corps. Since participants in the events were responsible for learning the lessons, problematic aspects were selectively and intentionally overlooked for bureaucratic-political reasons, and unpleasant results often bypassed.[259] Without the theoretical background, Israeli officers frequently had no frame of reference for analyzing their combat conduct and resorted to the "bottom line test"—if the operational results were successful, the process was considered successful as well.[260] Ethnocentric and short on curiosity about surrounding cultures,[261] Israelis were relatively indifferent to foreign combat practice,[262] believing that "their" campaigns are peculiar, so that there is nothing to learn from others. They focused mostly on the technological aspects of (usually) American experience.[263] This weakness of learning lessons was particularly important with regard to envisioning future wars. Some proposals were made by Isaac Ben-Israel to establish the didactic learning process, but they were never implemented.[264] For most of its history, the IDF lacked a group of "philosophers" that could inquire into the future and envision the changes in military regimes.[265]

The Role of the General Staff

Given the weak planning staff in the Ministry of Defense, in the Knesset's foreign and security affairs committee, in the Ministry of Foreign Affairs, and in the National Security Council (NSC), the GS became Israel's only body capable of systematic staff work, strategy development, and implementation.[266] In fact, it functions more as a national security council than as a classic general staff.[267] As a by-product of this anomaly, the senior military leadership, which is involved in a wide spectrum of political-military activities, had less time to dedicate itself to classic military roles, such as crafting a "new theory of victory."[268] The general disrespect for military thought, multiplied by the overload imposed by various nonmilitary missions,[269] resulted in weak institutional commitment to the areas of military science and doctrine.[270] Compared to the US JCS, the IDF GS was highly centralized, at the expense of services and combat commands. However, if judged against the Soviet GS, it was very weak as an intellectual center of gravity. The strongest institution in the Israeli defense community, the IDF GS was far from being the "brain of the military."[271]

The IDF never developed specialization and professional training for its GS officers. There was no attempt to produce military experts with an interdisciplinary, synthetic grasp of warfare, deep knowledge of military art, or the ability to think in terms of supra-services. No career path was ever defined for the GS officers, nor was a specialized training program designed.[272] Officers possessed strong parochial loyalty to the corps that posted them to the position, and bureaucratic wars between different branches of the GS, which saw themselves as independent services, were commonplace. Decision making represented the bureaucratic bargaining among different GS factions, where branches presented their organizational goals as national security interests.[273]

The IDF operated in an absence of long-term conceptual and doctrinal vision. The Doctrine and Training Unit of the GS (TOHAD) concentrated chiefly on fragmented details of the professional corps' concepts of operations on techno-tactical levels. It did not develop a doctrine for integrated warfare or combined arms cooperation and did not ensure the assimilation of such doctrines through proper training. Improvisation frequently was a substitute for professionalism of staff officers. Under the anti-intellectual, pronounced aversion to abstract thinking, GS tended to reduce complex strategic problems "to the level of discrete, technical puzzles that must be solved quickly because of the pressure of events on the ground." Problems were resolved in an isolated and sequential manner as if they were not interconnected. The comprehensive view and thorough future calculus were neglected in favor of a pragmatic but narrow technical perspective and ad hoc day-to-day considerations. Comprehensiveness, strategy, and long-term consideration, expected from the GS, were lost in such a process.[274]

CONCLUSION: CULTURAL FACTORS AND THE ISRAELI RMA

Israel acquired the RMA arsenal more by default than by design. Driven by obsessive siege mentality, and a quest for absolute security, Israel established an ultrasophisticated and innovative defense industry. Fixating on preserving and increasing its qualitative edge vis-à-vis neighboring militaries enabled it to develop in certain areas a technological arsenal so advanced that the superpowers did not possess, let alone deploy, even a fraction of it. Although initially the Israeli way of war was based on a qualitative edge consisting of a synergy between creative stratagems, motivation, unit cohesion, and superior personal qualities of its soldiers, after the early 1970s the IDF turned to technological solutions and to sophisticated weaponry as a cure-all in military affairs. This metamorphosis in interpretation of qualitative edge in favor of growing reliance on technology encouraged the neglect of many nonmaterial dimensions and resulted in incremental debilitation of Israeli military thought.

The IDF innovations were usually techno-tactical and sectarian, and resulted from a synergy among canny stratagem, advanced technology, and assertiveness on the battlefield. Innovations were rarely the product of thorough deductive thinking and occurred as creative tactical improvisations by assertive officers of a particular service; they did not produce subsequent transformation of the IDF ways of war. Technologies and weaponry from summer 1982 were further developed but remained a sectarian, tactical modernization of the IAF, with no significant impact on the concepts of operation of other corps, let alone GS's vision of future war. The new weaponry was developed and procured but was treated only as a force multiplier and incorporated into the existing force structure and concept of operations, without changes in the architecture of Israeli military power. Offensive romanticism was so deeply ingrained in the IDF operational mind-set that it prevented acknowledging the impact of new weaponry on the offense–defense balance and reexamining the posture of Israeli military doctrine. Short-term orientation bounded the horizons of Israeli military thought in general and conditioned the Israeli approach to the RMA in particular. Vigorous technological innovations and creative ad hoc improvisations coexisted in the IDF with conceptual conservatism. A pioneer in technology implementation, the IDF did not produce any significant leap forward in its vision of the future war. The Israeli approach to transforming the IDF along RMA lines was profoundly influenced by the US Defense Transformation. The application of American ideas, however, was frequently implemented in a mechanical manner, because the IDF beyond OTRI was mostly focused on the technological attributes of the innovation and was less disposed to considering conceptual ones.

An additional factor that fostered the gap between theoretical and practical matters in the IDF was a preference for promoting practitioners over theoreticians. In keeping with the attributes of the practical, assertive, socially constructed "new kind of Jew," the anti-academic and anti-intellectual tradition in the IDF placed obstacles in the way of abstract theoretical thinking. As a consequence of the IDF's emphasis on practice over learning and its disinclination to study modern war in a multidimensional manner, it became conservative, even inept in the conceptual field. With no formal professional education, IDF officers thought and operated in tactical terms, concentrating on giving ad hoc, piecemeal solutions to immediate problems.

The overload of the GS with nonmilitary missions, compounded by a general Israeli underestimation of the value of military science, resulted in a weak institutional commitment to the fields of military theory and doctrine. Its central position notwithstanding, the IDF GS lacked experts with an interdisciplinary, synthetic grasp of warfare, profound knowledge of military art, and an ability to think in supra-service terms. The IDF operated without long-term conceptual and doctrinal vision. Under the anti-intellectual, declared aversion to abstract thinking, GS tended to reduce complex strategic problems to the level of technical puzzles that must be solved quickly because of time pressures. Each event was turned into a separate issue to be discussed ad hoc, without reference to its connection to other events. The Doctrine and Training Unit of the GS mostly concentrated on the fragmented details of the techno-tactical concepts of the services. In keeping with the Israeli tradition of authority delegation, most of the conceptual innovations originated in a bottom-up manner, from the services. The comprehensive view and thorough future calculus were neglected in favor of a pragmatic but narrow technical perspective and day-to-day, parochial considerations.

That preoccupation with routine and ever-pressing missions left military leadership little time to think about the future of warfare or to experiment with new concepts. It predisposed the IDF away from posing RMA-type questions on paradigm shifts in the nature of war. Most of its intellectual energy was invested in assessments of "where are we?" rather than analysis of "where are we going?" The most damaging consequence of this tendency was a fundamental inability to anticipate and adapt to changing military regimes. For most of its history, the IDF lacked a group of educated officers willing to think in a broad way about the nature of warfare. The tactical smartness of the IDF was not translated into an intellectual inclination to make qualitative leaps in military thought. Although academics and civilian experts engaged in the conceptual debate about the changing nature of warfare after the early 1990s, their overall impact on the defense establishment was inconsequential. As an institution, the IDF made only piecemeal, intermittent attempts to comprehend the emerging security regime and to transform itself accordingly. The prevailing intellectual conservatism and lack of relevant organizational framework hampered it from putting these concepts together in a coherent form and translating abstract theoretical ideas into concrete military reform.

Acknowledgment and transformation along RMA lines began to gather momentum when a group of reformers established itself in the IDF. However, the postmodern, conceptual odyssey that these reformer-theoreticians initiated, and which spread the RMA ideas, became a problematic conceptual experience for the IDF. Their use of dense, jargon-filled language and an uncritical application of postmodern disciplines to military affairs, especially given the weak theoretical foundations of the IDF commanders, produced confusing results. The IDF lacked sufficient intellectual capital to digest these ideas and to produce the theoretical antithesis in order to engage these new concepts critically. Although OTRI became the knowledge authority for the IDF, under its auspices Israeli military thought became so complicated that in its damaging influence, it even dwarfed the traditional anti-intellectualism.

The transformation attempts resulted in serious cognitive dissonance inside the IDF. Practical and pragmatic Israelis were more at home with a logical-analytical approach to processing data than with a holistic-dialectical grasp of reality, which favors descriptive knowledge. The orientation toward doing, coupled with the trait of anti-intellectualism, led Israelis to value an inductive approach, where praxis and procedural knowledge precede long-range vision. Having an acute sense of practicality, assertiveness, and a tendency to improvise, underlain by low-context orientation, it is not surprising that the mainstream orientation of the Israeli mentality favors procedural knowledge over the descriptive type. The recent conceptual transformation in the IDF was in fact an attempt to implant a holistic-dialectical way of thinking into an organization whose mainstream was strictly logical-analytical. Practical and pragmatic Israelis in general, and IDF in particular, were strongly oriented toward procedural data, while in the case under discussion an attempt was made to reorient the IDF toward developing descriptive knowledge. This endeavor rubs against the grain of Israeli strategic culture.

The aspirations of the Israeli RMA architects to take a big step forward in the development of military thought resulted in a massive conceptual cacophony in the IDF. Resistance to conceptual changes, poor professional military education, an inability to engage new ideas critically and constructively, and the doctrinal confusion that originated in the IDF toward and during the 2006 war, perfectly reflect most of the traits in Israeli strategic culture and its military tradition of the *balagan meurgan*—"organized mess." When the IDF entered into the Lebanon war in 2006, it was in the transition period between the *crisis of knowledge* and the *emergence of the new paradigm,* using Kuhn's terms. This period of revolutionary transition is, according to Kuhn, neither rapid nor smooth. Nonetheless, the new CONOP, which represented the quintessence of Israeli views on the nature of warfare in the RMA era, constituted a revolution in itself. Indeed, it was the first time that the IDF had codified in written form a field manual for the operational and strategic levels of war. Paradoxically, the capabilities for this new theory of victory had existed for more than a decade, but Israeli military thought has not kept pace with the sophistication of IDF equipment.

CONCLUSION

CULTURAL FACTORS AND MILITARY INNOVATION: COMPARISON OF THE USSR, THE US, AND ISRAEL

This book addressed the puzzling intellectual history of the transformation in warfare in the USSR, the US, and Israel to show the impact of sociocultural characteristics on military innovations and on receptivity toward new military practices from abroad. The fact that the Soviet "new theory of victory" chronologically preceded technological procurement and combat experience, while in the American and the Israeli cases the cultivation of the technological seeds preceded the maturation of the conceptual ones, demonstrated that the RMA originated differently in diverse ideational milieus. This research has assumed that cultural factors intervened in the causal mechanism between technology and innovation, conditioning the intellectual paths taken toward the same paradigmatic change in military affairs. The book has illustrated how variance in strategic cultures accounts for the various ways in which military innovations, based on similar technologies, develop in different states.

The Russian military's traditional lack of fascination with technology favored a broader conceptualization of the relationship between advanced weapons and defense transformations. In the atmosphere of technological romanticism, which prevailed in the US, mostly functional and tactical application of the advanced technologies took center stage. The proactive, pragmatic, and practical American strategic personality channeled the innovation to satisfy specific requirements related to a narrow set of operational threats and possessed only intuitive understanding of the revolution that was about to occur. Israelis, in a vein similar to that of their American ally, acquired the RMA arsenal more by default than by design. Neglecting nonmaterial aspects of innovation in favor of a technological edge, the IDF developed and procured new weaponry but treated it as a force multiplier for the existing concept of operations, without changing the architecture of Israeli military power. The innovative utilization of the new weaponry was nothing more than successful but sectarian tactical improvisation.

Institutionalized military science in the USSR created an intellectual atmosphere that fostered attention to discontinuities in military affairs. In keeping with the authoritarian Russian mind-set, the General Staff executed the highest intellectual authority. The role given to the GS in the Soviet military system, the professional approach of the GS officers, and the intellectual atmosphere in its directorates enabled the "brain of the army" to capitalize on its intellectual potential and to develop constantly fresh military ideas about the future of war. When the new technologies arrived in the mid-1970s, the Soviets already possessed the scientific framework and the advanced system of terminology to acknowledge the emerging discontinuity. Americans and Israelis were less prepared institutionally to diagnose Revolutions in Military Affairs. In keeping with decentralized liberal culture, conceptual and organizational military innovations originated in the US in a bottom-up manner, from the services, and not top-down from the Joint Chiefs of Staff or the DoD. The American JCS had no ethos of being a "brain of the military," and consequently strategic and long-term defense planning was weakly institutionalized there. The JCS lacked a powerful bureaucracy that could produce an effective cross-service vision that would affect the long-term development of US military power. The chairman was a budgetary manager and occasional operational planner, but not a deductive thinker about the nature of war. He remained the captive of the services' parochialism and lacked the intellectual capital to generate deep, crosscutting, long-term observations. The overload of the Israeli GS with nonmilitary missions, multiplied by the general lack of Israeli respect for military science, resulted in a weak institutional commitment to the areas of military theory and doctrine. Its central position notwithstanding, the Israeli Defense Forces GS was hardly a "brain of the military," lacking experts with an interdisciplinary, synthetic grasp of warfare, a deep knowledge of military art, and the ability to think in supra-service terms.

The intellectual atmosphere established in Soviet professional circles fostered the diagnostics of discontinuities in the nature of war. Early in Soviet history, the Bolsheviks sought to establish a unique military science and a systematic framework of analysis that would provide a master key to military reality. The Soviet military developed comprehensive methodological guidelines on how to think about war scientifically. As their most fundamental task, military theoreticians were obliged to envision the nature of future war and to learn to identify the indications of emerging discontinuities in military affairs. The US military was far less prepared for grasping the RMA. For generations, a battle of annihilation and enemy attrition by superior firepower had been the American way of war. This industrial approach to warfare accounts for the relative disfavor of the American theoretical military tradition. One of the corollaries of this approach was that the notion of operational art as a theoretical concept was rejected by the US military until the 1980s. The aim of "collapsing the enemy" by operational maneuver rather than simply annihilating it with firepower conceptually diverged from the established American strategic tradition. Because ideas about paradigmatic changes in the nature of war originate on the operational level of military thought, the absence of this intellectual layer was a serious obstacle that prevented American strategists from thinking in terms of the RMA. For Israelis to diagnose the RMA was even more difficult. It was practitioners rather than theoreticians who reached the military's upper echelons. In keeping with the ethos of being practical and assertive, the anti-academic and anti-intellectual tradition in the IDF decreased the capacity for abstract theoretical thinking. Because the IDF put practice before learning and was disinclined to study war in a multidimensional manner, it became conservative and ineffectual in the conceptual field.

The impact of cognitive style also played an important role in anticipating and acknowledging the RMA. The traditional Russian cultural predisposition to holistic inferential procedures, compounded by Soviet dialectical postulates, fostered synthetic thinking by Soviet military theoreticians. The Soviet experts, in keeping with the Russian polychronic, holistic culture, saw problems as interrelated and required a generalized frame of reference. The achievements of scientific-technological progress represented the *focal point*, while the changes in the ways and means of military operations represented the *field*. The introduction of weaponry, communications, and intelligence sensors associated with microprocessors swiftly became organically connected in the Soviet mind with this technology's impact on the nature of war. This way of thinking made it easier to conclude whether the new technology constituted the continuation of an existing trend in military art or had brought a revolutionary transformation in military affairs.

The American cultural predisposition to a logical-analytical cognitive style made the US military less receptive to the kind of reasoning needed for perceiving and comprehending dialectical transformations in military affairs. In forecasting attempts, the logical-analytical style directed American mental formation toward focalism—a tendency to focus attention narrowly on the upcoming event and to detach it from its overall implications. During the introduction of PGM weaponry into the battlefield, the mainstream of American military experts concentrated on the *focal point*—mechanical application of the new technologies on the tactical level—and detached it from the *context*—the implications of this new weaponry on the ways and means of waging operations. The concentration on the focal technologies at the expense of the broader contextual implications diverted the US military from swift comprehension of the paradigmatic change in the nature of war.

A similar cognitive orientation of the Israelis brought them to the same place. Practical and pragmatic, Israelis were more at home with a logical-analytical approach to processing data than with a holistic-dialectical grasp of reality, which favors descriptive knowledge. The orientation toward doing, coupled with the trait of anti-intellectualism, caused Israelis to value an inductive approach, in which praxis and procedural knowledge precede deep visions. With an acute sense of assertiveness, and a tendency to improvise, underscored by a low-context orientation, it is not surprising that the mainstream Israeli thinking favored procedural knowledge over the descriptive one, and subsequently became less receptive to acknowledging paradigmatic change in the nature of warfare.

The pragmatic American approach had the advantage of efficacy and quality, and Israelis excelled in assertiveness and creative improvisations; however, both, being strongly inclined to procedural knowledge, suffered from the disadvantage of superficiality, short-range vision, and piecemeal forecasting efforts. Although the Soviet approach had the advantage of powerful deep thought, a "being and not a doing culture," with a predisposition to declarative knowledge, the Soviets were traditionally good at theorizing, but pathologically bad at connecting words and deeds. The sophisticated Soviet MTR concepts were incompatible with the country's capacity to accomplish them.

The intellectual history of this innovation provides a striking example of adaptive learning. The cases presented here have demonstrated three different trajectories of diffusion of RMA: starting in the late 1970s, the Soviets used American technology as a frame of reference and produced an innovative conceptual envelope around it; starting in the late 1980s, the US imported Soviet concepts, developed them further, and conveyed them to the Israelis, who have adopted and used the described combat systems since 1982. The Soviet tradition of military futurology not only influenced American and Israeli paths to the IT-RMA but also shaped methodologies that guided the development of the "new theories of victory" in two countries. Since the late 1990s, the Israeli approach to developing military thought has become an amalgamation of Soviet and American theories. Cultural factors played an important role by affecting the adopting state's interpretation and absorption of revolutionary innovations: an inclination toward ethnocentrism resulted in "mirror imaging," which prevented American analysts from understanding Soviet theoretical writings for about a decade; Israelis traditionally concentrated on material factors rather than conceptual lessons and frequently adopted American ideas in a mechanical and uncritical manner, focusing mostly on technological aspects of the innovation. Moreover, inspired by foreign (Soviet and American) professional ideas, Israelis introduced approaches that did not necessarily apply to their native operational and cognitive environment.

"Norm entrepreneurs" or "epistemic agents" frequently become the chief enablers of innovative changes within their national professional communities. They interpret events, frame the discourse, and construct new consensuses.[1] Studies in military innovations suggest that entrepreneurs' proximity to power—be they organizations or influential individuals—is critical to success.[2] The intellectual history of the RMA in the three countries fully supports Stephen Rosen's argument that successful entrepreneurs of military innovations are visionary leaders, who mobilize change within their own organizations.[3] The professional discourse that interpreted the emerging technologies as a revolutionary development was framed in each of the three countries to a large extent by "norm entrepreneurs" who occupied influential bureaucratic positions in the military or defense establishments: in the Soviet case the General Staff, the Office of Net Assessment in the American case, and OTRI in the case of the Israeli RMA transformation.

To what extent is it possible to generalize the findings of this work and apply them to other fields of strategic behavior in given countries? This book focused on conventional innovations and has not discussed the way in which cultural factors shaped nuclear doctrines in the USSR, the US, and Israel.

Testing the findings of this research against empirical evidence from the nuclear field and tracing the cultural roots of the nuclear doctrines in the three countries could verify the credibility of this book's argument. To accomplish this important task would be another book-length project. Here it is sufficient to say that the intellectual history of the nuclear revolution demonstrates similar variations in approaches to innovation across the cases.

Almost a decade after the first explosion, Eisenhower's "New Look" in the mid-1950s "reflected the belief that nuclear weapons had revolutionized warfare." However, the roles of the services were not redefined according to the necessities of the nuclear age until the late 1950s. Nuclear strategy emerged as a separate discipline only toward the mid-1950s, and until then many in the US defense and military establishment treated atomic weapons as conventional firepower. It took the US military establishment more than a decade to conceptualize consequences of this discontinuity in the nature of war. An understanding of the revolutionary nature of the weapons emerged more slowly than the maturation and sophistication of nuclear capabilities.[4] Although the Soviet military initially demonstrated a somewhat similar, "conventional" attitude to nuclear capabilities, the Russian approach was different. The Soviets started to conceptualize the emerging nuclear weapons and thinking through conceptual implications, simultaneous with attempts to acquire atoms through intelligence collection. Even before the arsenal and means of delivery became more sophisticated, the emerging capabilities were diagnosed by the Soviet military theoreticians as the MTR. This conceptual verdict obliged them to think through the impact of the new capabilities on the character and conduct of warfare, in terms of both concepts of operations and force structure. As in the case of the IT-RMA, the cultivation of the conceptual seeds preceded the maturation of the actual capability.[5]

Israel initiated its nuclear project in the 1950s, without carefully analyzing the long-term strategic objectives and military implications of the program. Most debates focused on the immediate issues related to building the capabilities and infrastructure but avoided articulating the complex doctrinal issues concerning nuclear weapons. Despite very little strategic guidance, a rudimentary nuclear capability matured on the eve of 1967, and the improvised preliminary contingency plans were drafted by the leaders of the Israeli nuclear project, not in response to any specific political or military request. In May 1967 the "technologists" pushed Israel over the nuclear threshold, while the "politicians" were far from outlining a clear nuclear strategy. Similarly, during the 1973 war, in the midst of conventional contingency, improvised nuclear signaling took place not according to any articulated doctrinal concept. After some ad hoc considerations made in the narrow circle of decision makers, Israel utilized its capability as a coercive tool. Only in subsequent years did Israel start to articulate a conceptual framework to envelope the existing capabilities within a kind of a doctrine. "Israeli thinking and planning about the use of nuclear weapons evolved more slowly than its nuclear infrastructure," argues Avner Cohen.[6]

IMPLICATIONS FOR SCHOLARSHIP AND POLICY

Findings from this work reveal important implications for the scholars of security, military innovation, and strategic cultural studies and for the policy planners in defense and military establishments. I focus here on several vistas of future research related to conventional, subconventional, and nonconventional issues relevant for both academics and practitioners.

Whether the transformation in the nature of warfare discussed in this book represented a revolutionary or an evolutionary discontinuity in military affairs is one of the central questions of current security studies. However, that inquiry has intentionally been left outside the scope of this work for theoretical and historical reasons. From a theoretical perspective, this research did not aspire to assess the effectiveness of the IT-RMA-type warfare in modern military conflict. It primarily focused on understanding the intellectual dynamic by which defense experts conclude that paradigmatic changes in the nature of warfare are under way, regardless of the ultimate effects of those changes. The main goal of this research was to depict the intellectual history of the innovation without judging its effectiveness, and to discuss the imprint of cultural attributes on developing the various versions of the RMA. Even if the "new theory of victory" were to have found itself on the wrong side of history, the chronicles of ideas would have been of greater interest to me than the final result.

The second reason is historical; it is important to remember that in all three countries, RMA ideas were born in the context of conventional and nuclear Cold War contingencies. Beyond all doubt, MTR and RMA ideas constituted the most revolutionary development in the military thought of the final stages of the Cold War. However, it would be a mistake to judge this Cold War theory of victory in the context of the post-9/11 environment. Andrew Marshall, whose ideas fostered the diffusion of the RMA discourse within official military circles throughout the world, was one of the first to claim that the IT-RMA had very little application to certain kinds of conflict.[7] Indeed, in all three countries the RMA type of war showed itself to be of limited utility in large-scale counterinsurgency campaigns: the Soviets experienced the irrelevancy of the MTR concepts in Afghanistan, the US military has been operating below its own expectation in Iraq, and the IDF failed to generate swift military victory using RMA techniques in 2006 in Lebanon.

All three militaries found themselves on battlefields not consistent with the visions of the idealized RMA. Eventually, they were forced to significantly modify their visions of war and adjust them to the current security environment. H. R. McMaster showed how fascination with military futurology resulted in wishful thinking, when the proponents of the IT-RMA outlined how they "would like to fight and then assumed that the preference was relevant." This self-delusion, of course, corrupted the ability to think critically about warfare.[8] However, these findings support the argument by Watts and Murray that military institutions need to constantly invest significant intellectual energy in developing a vision of a future war. Because uncertainty cannot be totally eliminated, this vision is almost certain to be vague and incomplete, rather than detailed, precise, or predictive.[9] The fact that oracle forecasts in military affairs are problematic should not dissuade analysts from modest but energetic intellectual endeavors. Although what Secretary of Defense Robert Gates termed "next-war-icity" might turn into a dangerous professional pathology,[10] preparations for "this war" and thinking about it have to start before "this war" actually needs to be fought. Otherwise, a military organization might become irrelevant for the current security environment, as H. R. McMaster has outlined in his works. Lack of vision about continuities and changes in the nature of war is as dangerous as wishful thinking about it. Without the intellectual-institutional commitment to craft a vision of future war and the simultaneous commitment to constantly and critically engage the new paradigm, the military organization will almost certainly lose its relevance for the given security regime.

Although many defense experts focus today on counterinsurgency and irregular warfare, conventional military innovations still matter very much. As Michael Horowitz and Dan Shalmon argue, a change in the correlation of forces brought about by innovation in specific conventional domains may put the traditional superiority of the state at a higher risk of danger than that brought about by asymmetrical challenge. The state-on-state "knockout blow" might be a less likely menace but is a much more grave threat for the state-actors of international security.[11] Today, the conventional military forces of several big status quo and rising powers are in the midst of vigorous defense modernization reforms. Application of the analytical approach outlined in this book can help to anticipate directions in which these actors might channel their innovative ideas. This approach could clarify some aspects of current military transformations elsewhere that seem puzzling and counterintuitive to Western observers. For example, consider the Chinese foreign and defense policy and its vigorous force modernization, which have recently attracted significant scholarly attention. As the works of Jacqueline Newmyer demonstrate, analyzing Chinese military modernization through an ideational-cultural lens—seeing China through Beijing's eyes—significantly improves the ability to explore, explain, and anticipate the directions of the People's Liberation Army's (PLA's) force buildup.[12]

Analytical traps confront intelligence and defense analysts who try to uncover foreign ways of war. Although it is difficult to know oneself in terms of strategic culture,[13] it is very important to attempt such self-reflection. The concept of "intelligence culture"—a recent outgrowth of "strategic culture theories"—not only helps to better understand the other side but also guides investigation of how one's own strategic culture conditions assessments of foreign military innovation. Recent works by Uri Bar-Joseph, Rose McDermott, and other scholars[14] show that cultural factors significantly influence traits in data collection and impact on the quality of information processing. Utilizing the concept of "intelligence culture," scholars explain the variance in the predisposition to analytical pathologies across intelligence institutions. It also illuminates why certain organizations are more likely to fall victim to specific analytical obstacles than others. The impact of cultural factors on the behavior of intelligence organizations is the most promising avenue of academic research. It has an immediate benefit to practitioners engaged in organizational reforms, investigation of analytical failures, and improvement of the everyday performance of intelligence analysts.

Cultural analysis has immediate applicability to counterinsurgency and irregular warfare as well. We know very little about ideational factors that condition strategic choices and inform the operational behavior of the Salafi jihadists.[15] Our understating of jihadi strategic culture is pedestrian when compared to our knowledge of the ideational factors that underlie security communities elsewhere. We are familiar with the ideational foundations of the Islamic way of war, but less so with operational military norms and patterns of thinking and action as cultivated by the jihadi narrative. These norms, however, strongly influence operational behavior, the theory of victory, and way of war of mujahedeen. Knowledge of Islamic strategic culture might be helpful in identifying common denominators for patterns of behavior and thought among jihadists, regardless of their geographical deployment or organizational affiliation. Although mujahedeen possess a range of views on military affairs, it might be possible to outline some conceptual similarities and continuities across several jihadi entities. Since strategic culture is a repertoire of ideas and options, rather than a determinant of behavior, it has its explanatory boundaries and should be handled with care. Patrick Porter's *Military Orientalism*[16] makes reasonable precautions about culturally paternalistic and insensitive generalizations about the Islamic way of war. Being aware of this precaution, an exploration of the cultural roots of jihadi strategic behavior is an important research topic to engage.

On a more theoretical side, the question of how cognitive style as part of strategic culture conditions thinking about military affairs is another promising vista of research. This book tried to identify how an intellectual approach to the generation of the RMA was shaped by cognitive styles of the defense communities in the three countries under study. Although process-tracing did not establish strong causation, the portions of this book dealing with this issue have identified a new phenomenon and demonstrated obvious variance across the cases. If developed further, this analytical approach might be a potentially useful tool for portraying and explaining military innovations. Additional theoretical imports to security studies from cultural-cognitive psychology and a deeper look into the link between culture and cognition in security affairs, based on fresh empirical data, will clarify whether there is a real or an imagined elephant in the room.

As new powers in the Middle East and Asia are likely to join the nuclear club, the world is entering a "second nuclear age" in which nuclear challenges might differ significantly from the ones of the (post–) Cold War era. What would be the patterns of strategic behavior of the actors who have recently acquired nuclear weapons? What security doctrines will newcomers to the nuclear club tailor to their newly acquired capabilities? How will they communicate their nuclear intentions and capabilities to other states? How will other states be able to identify strategic signals of the newcomers? How to make sure that the newcomers know to interpret signals and bargaining techniques of other states? The ability to diagnose the nature of strategic interactions of what Stephen Rosen defines as "the post-proliferation era" is of unquestionable importance. Learning through historical analogies of the Cold War might not be very helpful. Since rational behavior is neither objective nor universal, but subjective and culturally dependent, it is doubtful that the superpowers' logic can be extrapolated to the strategic conduct of the newcomers to the nuclear club. To reduce uncertainties about how the newcomers might approach their newly acquired nuclear power, one should adopt a less rational-mechanical, but more idiosyncratic, approach. A systematic application of cultural analysis might improve an ability to understand what factors and considerations will condition countries' attitudes toward nuclear power capabilities of their own and the same capabilities in the hands of others. Research assumptions that are not in harmony with the strategic culture of the analyzed country might result in serious miscalculations when dealing with questions of strategic signaling, deterrence, and nuclear weapons use.[17]

This book does not contribute to the practitioners by ascertaining the most effective military doctrine from the assortment available today. Nevertheless, the nuanced analysis of the recent RMA presented in this book suggests some lessons for security experts looking for a new theory of victory. The intellectual history of the RMA in the US, the USSR, and Israel makes a compelling case for the need for military theory to lead technological development and weapons procurement or, at least, to coevolve simultaneously. A self-satisfied, intellectually stagnant belief in possession of a technological fairy stick can be damaging to any military organization.[18] Side by side with the danger of technological determinism is an equal risk of ignoring the effects of technology and thinking that either enduring bodies or creative improvisation can overcome all obstacles.[19] The ideal type of military innovation might well be a synthesis of the Russian-Soviet deep theoretical grasp of the nature of warfare and military art, American technological genius in weaponry development and production, and Israeli operational assertiveness and creative improvisation on the battlefield. The presence of this triad could tremendously increase the chances of any military organization to embark upon a relevant military buildup, to prepare for the next, rather than the last, war, and to achieve victory in haste.

Elaborating on the intellectual history of the RMA, this book has illuminated how under the impact of cultural factors countries interpret differently the changing character of war. In the future, both state and nonstate actors will continue to develop military knowledge, and security experts will continue to uncover foreign military innovations. In each case there will be a need to figure out the tools of war ("hardware") and anticipate their application ("software").[20] The task with regard to "software" will be much more demanding, and a cultural approach will be indispensable for it. Ideational analysis does not unconditionally predict the strategic or operational behavior of international security actors. However, it offers a smart, systematic, and thorough way to think about it. As such, a cultural approach to the issues of strategy better prepares security experts from academia and from the world of praxis to think about the unthinkable and expect the unexpected.

NOTES

NOTES

Introduction

1. Michael Horowitz and Stephen Rosen, "Evolution or Revolution?" *Journal of Strategic Studies* 3 (June 2005). The initial definition was offered by Andrew Krepinevich, "Cavalry to Computer: The Pattern of Military Revolutions," *National Interest*, no. 37 (Fall 1994); Colin Gray, *Recognizing and Understanding Revolutionary Change in Warfare* (Carlisle, PA: US Army War College, 2006), p. vi. For the impact of anthropological, demographic, natural, physiological, and economic factors on the transformation of warfare, see Azar Gat, *War in Human Civilization* (Oxford: Oxford UP, 2006); Stephen Rosen, *War and Human Nature* (Princeton, NJ: Princeton UP, 2006). For an overview of the RMA, see Project on Defense Alternatives, The RMA Debate, http://www.comw.org/rma/.

2. Andrew W. Marshall, director of Net Assessment, Office of the Secretary of Defense, *Statement for the Subcommittee on Acquisition and Technology* (Washington, DC: Senate Armed Services Committee, 5 May 1995).

3. Gray, *Recognizing and Understanding*, p. 11. For the impact of the most significant RMAs in modern history—such as the gunpowder revolution, the first industrial revolution, the second industrial revolution, and the information age revolution—on the nature of warfare, see Max Boot, *War Made New: Technology, Warfare and the Course of History, 1500 to Today* (New York: Gotham Books, 2006).

4. Boot, *War Made New*, p. 10.

5. Barry Watts, *What Is the Revolution in Military Affairs?* (Northrop Grumman Analysis Center, 6 April 1995), pp. 5–6; Richard O. Hundley, *Past Revolutions, Future Transformation: What Can the History of Revolutions in Military Affairs Tell Us about Transforming the US Military?* (Santa Monica, CA: RAND, 1999), p. 27.

6. Earl H. Tilford, *The Revolution in Military Affairs: Problems and Cautions* (Carlisle, PA: Strategic Studies Institute, US Army War College, 1995), p. iii.

7. Marshall, *Statement for the Subcommittee*, pp. 3–5; Watts, *What Is the RMA?* pp. 5–6. The necessity to discern RMA goes beyond the demands of battlefield effectiveness and is essential for the whole process of defense management. When there is no solid vision of a future war, or without a coherent military doctrine and concepts of operation, it is highly probable that military organizations will build and equip their forces not in compliance with the demanding strategic environment.

8. Hundley, *Past Revolutions*, p. 13.

9. Barry Watts and Williamson Murray, "Military Innovation in Peacetime," in Williamson Murray and Allan Millet, *Military Innovation in the Interwar Period* (New York: Cambridge UP, 1996), pp. 404–6.

10. Gray, *Recognizing and Understanding*, p. 3.

11. Theodor W. Galdi, *Revolution in Military Affairs?* (CRS 951170F, 11 December 1995), p. 3.

12. The term "new theory of victory" is borrowed from Stephen Peter Rosen, *Winning the Next War: Innovation and the Modern Military* (Ithaca, NY: Cornell UP, 1994), p. 20.

13. Michael G. Vickers and Robert C. Martinage, *The Revolution in War* (Washington, DC: Center for Strategic and Budgetary Assessments, 2004), p. 7.

14. Eliot A. Cohen, "Change and Transformation in Military Affairs," *Journal of Strategic Studies* 27, no. 3 (September 2004); Avi Kober, "Does the Iraq War Reflect a Phase Change in Warfare?" *Defense and Security Analyses* 21, no. 2 (June 2005), pp. 121–42.

15. Vickers and Martinage, *Revolution in War*, p. 14.

16. Chris C. Demchak, "Creating the Enemy: Global Diffusion of the Information Technology-Based Military Model," in *Diffusion of Military Technology and Ideas*, ed. Emily O. Goldman and Leslie C. Eliason (Stanford, CA: Stanford UP, 2003), pp. 307–10.

17. Although the key inventors of the machine gun were the Americans, the European armies in Africa were the first to use them. Although the British invented the tank, the Soviets were the first to understand and the Germans were the first to show how to fully exploit its potential. Although the British conducted the first carrier air raid, the American and the Japanese invented carrier warfare. Hundley, *Past Revolutions*, pp. 14–15.

18. Andrew W. Marshall, "Foreword," in Goldman and Eliason, *Diffusion of Military Technology*, p. xiv; Andrew Marshall, *Some Thoughts on Military Revolutions* (Memorandum for the Record: Office of the Secretary of Defense, 23 August 1993).

19. Jeffrey S. Lantis, "Strategic Culture: From Clausewitz to Constructivism," *Strategic Insights* 4, no. 10 (October 2005); Brice F. Harris, *America, Technology and Strategic Culture* (New York: Routledge, 2008), pp. 30–39.

20. Works from the first generation include: Jack Snyder, *The Soviet Strategic Culture: Implications for Nuclear Options* (Santa Monica, CA: RAND, 1977); Ken Booth,

Strategy and Ethnocentrism (New York: Holmes and Meier, 1981); Colin Gray, "National Style in Strategy: The American Example," *International Security* 6, no. 2 (Fall 1981), pp. 35–37, *The Geopolitics of Super Power* (Lexington, KY: UP of Kentucky, 1988), pp. 42–43, and *War, Peace, and Victory: Strategy and Statecraft for the Next Century* (New York: Simon and Schuster, 1990); Carl Jacobson, *Strategic Power: USA/USSR* (New York: St. Martin's Press, 1990).

21. Yitzhak Klein, "A Theory of Strategic Culture," *Comparative Strategy* 10, no. 1 (1991), p. 3. See also Richard W. Wilson, *Compliance Ideologies: Rethinking Political Culture* (New York: Cambridge UP, 1992); Charles A. Kupchan, *The Vulnerability of Empire* (Ithaca, NY: Cornell UP, 1994).

22. Alastair Iain Johnston, "Thinking about Strategic Culture," *International Security* 19, no. 4 (Spring 1995), pp. 32–64, and "Cultural Realism and Strategy in Maoist China," in *The Culture of National Security: Norms and Identity in World Politics*, ed. Peter J. Katzenstein (New York: Columbia UP, 1996); Colin Gray, "Strategic Culture as Context: The First Generation of Theory Strikes Back," *Review of International Studies* 25 (1995).

23. Lantis, "Strategic Culture." Also see: Alexander Wendt, "Constructing International Politics," *International Security* 20, no. 1 (1995); Peter J. Katzenstein, Robert O. Keohane, and Stephen Krasner, "International Organization and the Study of World Politics," *International Organization* 52, no. 4 (1998); Ted Hopf, "The Promise of Constructivism in International Relations," *International Security* 23, no. 1 (Summer 1998), p. 914; Jeffrey W. Legro, "Culture and Preferences in the International Cooperation Two-Step," *American Political Science Review* 90, no. 1 (March 1996), pp. 118–37.

24. Peter J. Katzenstein, ed., *The Culture of National Security: Norms and Identity in World Politics* (New York: Columbia UP, 1996); Jeffrey W. Legro, *Cooperation under Fire: Anglo-German Restraint during World War II* (Ithaca, NY: Cornell UP, 1995); Alastair Iain Johnston, *Cultural Realism: Strategic Culture and Grand Strategy in Chinese History* (Princeton, NJ: Princeton UP, 1995); Stephen Peter Rosen, *Societies and Military Power: India and Its Armies* (Ithaca, NY: Cornell UP, 1996); Elizabeth Kier, *Imagining War: French and British Military Doctrine between the Wars* (Princeton, NJ: Princeton UP, 1997); Ken Booth and Russell Trood, eds., *Strategic Cultures in the Asia-Pacific Region* (London: Macmillan, 1999).

25. Valerie M. Hudson, ed., *Culture and Foreign Policy* (Boulder, CO: Lynne Rienner, 1997).

26. Colin Gray, *Out of the Wilderness: Prime Time for Strategic Culture* (Fort Belvoir, VA: Defense Threat Reduction Agency, 2006), p. 5; Ronald L. Jepperson, Peter J. Katzenstein, and Alexander Wendt, "Norms, Identity and Culture in National Security," in Katzenstein, *The Culture of National Security*, pp. 54–55; Peter Katzenstein, *Cultural Norms and National Security: Police and Military in Postwar Japan* (Ithica, NY: Cornell UP, 1998), pp. 17–20; Theo Farrell, *The Norms of War: Cultural Beliefs and Modern Conflict* (London: Lynne Rienner, 2005), *The Sources of Military Change: Culture, Politics,*

Technology (London: Lynne Rienner, 2002), "Transnational Norms and Military Development," *European Journal of International Relations* (2002), and "World Culture and Military Power," *Security Studies* (2005).

27. Stephen Peter Rosen, "Military Effectiveness: Why Society Matters," *International Security* 19, no. 4 (1995), pp. 5–31, and *Societies and Military Power*; Matthew Evangelista, *Innovation and the Arms Race: How the United States and the Soviet Union Develop New Military Technologies* (Ithaca, NY: Cornell UP, 1988); Thomas U. Berger, *Cultures of Antimilitarism: National Security in Germany and Japan* (Baltimore: Johns Hopkins UP, 1998); Thomas Banchoff, *The German Problem Transformed: Institutions, Politics, and Foreign Policy, 1945–1995* (Ann Arbor: U of Michigan P, 1999); Michael Eisenstadt and Kenneth Pollack, "Armies of Snow and Armies of Sand: The Impact of Soviet Military Doctrine on Arab Militaries," in Goldman and Eliason, *Diffusion of Military Technology*, pp. 63–93.

28. Theo Farrell, "Figuring Out Fighting Organizations: The New Organizational Analysis in Strategic Studies," *Journal of Strategic Studies* 19, no. 1 (1996), pp. 122–35.

29. Kimberly Marten Zisk, *Engaging the Enemy: Organization Theory and Soviet Military Innovation, 1955–1991* (Princeton, NJ: Princeton UP, 1993); Legro, *Cooperation under Fire*; Lynn Eden, *Whole World on Fire: Organizations, Knowledge, and Nuclear Weapons Devastation* (Ithaca, NY: Cornell UP, 2004); Deborah Avant, *Political Institutions and Military Change: Lessons from Peripheral Wars* (Ithaca, NY: Cornell UP, 1994).

30. Elizabeth Kier, "Culture and Military Doctrine: France between the Wars," *International Security* 19, no. 4 (Spring 1995), pp. 65–93; for cultural influences on doctrinal developments, see *Imagining War: French and British Military Doctrine between the Wars* (Princeton, NJ: Princeton UP, 1997); Alastair I. Johnston, *Cultural Realism: Strategic Culture and Grand Strategy in Chinese History* (Princeton NJ: Princeton UP, 1995); Thomas G. Mahnken, *Uncovering Ways of War: US Intelligence and Foreign Military Innovation, 1918–1941* (Ithaca, NY: Cornell UP, 2002).

31. Goldman and Eliason, *Diffusion of Military Technology*; Emily Goldman and Thomas Mahnken, *The Information Revolution in Military Affairs in Asia* (New York: Palgrave Macmillan, 2004).

32. Jeremy Black, *Rethinking Military History* (London: Routledge, 2004).

33. Jay Winter, *Sites of Memory, Sites of Mourning: The Great War in European Cultural History* (Cambridge: Cambridge UP, 1995); Manfred F. Boemeke, Roger Chickering, and Stig Foster, *Anticipating Total War: The German and American Experiences, 1987–1914* (Cambridge: Cambridge UP, 1999).

34. Catherine Merridale, *Night of Stone: Death and Memory in Russia* (Cambridge: Cambridge UP, 2001); Jeffrey Verhey, *The Spirit of 1914: Militarism, Myth, and Mobilizations in Germany* (Cambridge: Cambridge UP, 2000); Aaron L. Friedberg, *In the Shadow of the Garrison State: America's Anti-Statism and Its Cold War Strategy* (Princeton, NJ: Princeton UP, 2000).

35. Jay Winter and Emmanuel Sivan, *War and Remembrance in the Twentieth Century* (Cambridge: Cambridge UP, 2000); Anita Shapira, *Land and Power: The Zionist Resort to Force, 1881–1948* (Stanford, CA: Stanford UP, 1999); Vejas Gabriel Liulevicius, *War Land on the Eastern Front: Culture, National Identity and German Occupation in World War I* (Cambridge: Cambridge UP, 2000); Mark Johnston, *Fighting the Enemy: Australian Soldiers and Their Adversaries in World War II* (Cambridge: Cambridge UP, 2000).

36. Theo Farrell, "Memory, Imagination and War," *History* 87, no. 285 (2002), pp. 325–48; Craig M. Cameron, *American Samurai: Myth, Imagination, and the Conduct of Battle in the First Marine Division, 1941–1951* (Cambridge: Cambridge UP, 1994).

37. Trevor J. Pinch and Weibe E. Bijker, "The Social Construction of Facts and Artifacts—or How the Sociology of Science and the Sociology of Technology Might Benefit Each Other," *Social Studies of Science* 14, no. 3 (August 1984), pp. 399–441; Donald A. MacKenzie and Judy Wajcman, eds., *The Social Shaping of Technology* (Milton Keynes, UK: Open UP, 1999).

38. Donald A. MacKenzie, *Inventing Accuracy: An Historical Sociology of Nuclear Missile Guidance* (Cambridge, MA: MIT Press, 1993); Graham Spindary, *From Polaris to Trident: The Development of US Fleet Ballistic Missile Technology* (Cambridge: Cambridge UP, 1994).

39. Adam Grissom, "The Future of Military Innovation Studies," *Journal of Strategic Studies* 29, no. 5 (October 2006), pp. 905–34.

40. Jeffrey Larsen, *Comparative Strategic Cultures Curriculum: Assessing Strategic Culture as a Methodological Approach to Understanding WMD Decision-Making by States and Non-State Actors* (Prepared for Defense Threat Reduction Agency, 2006). For a comparison of all existing scholarly definitions of "strategic culture," see Lawrence Sondhause, *Strategic Culture and Ways of War: An Historical Overview* (London: Routledge, 2006), pp. 123–25.

41. Terry Terriff, "Innovate or Die: Organizational Culture and the Origins of Maneuver Warfare in the United States Marine Corps," *Journal of Strategic Studies* 29, no. 3 (October 2006), pp. 475–503; Thomas Mahnken, *Technology and the American Way of War* (New York: Columbia UP, 2008); Harris, *America, Technology and Strategic Culture.*

42. For the application of analytical frameworks and methodological practices from different disciplines to the research of strategic culture, see Jeannie L. Johnson and Jeffrey A. Larsen, "Comparative Strategic Cultures Syllabus" (Prepared for Defense Threat Reduction Agency, 2006), pp. 5–6.

43. Gray, *Out of the Wilderness*, p. 8. If all possible causal variables for state action are included, little space remains for explanations of behavior beyond strategic culture. Lantis, "Strategic Culture."

44. Ann Swidler, "Culture in Action: Symbols and Strategies," *American Sociological Review* 51 (April 1986), pp. 273–86.

45. For example, see Mahnken, *Technology and the American Way of War*, pp. 2–11.

46. Tilford, *Revolution in Military Affairs*, p. iii.

47. Ron St. Martin and Linda McCabe, *Implications of Culture and History on Military Development* (McLean, VA: SAIC, Prepared for OSD/Net Assessment, 1996), cited in Goldman and Eliason, *Diffusion of Military Technology*, p. 9.

48. Kenneth Waltz, *Theory of International Politics* (New York: McGraw-Hill, 1979).

49. Barry R. Posen, *The Sources of Military Doctrine: France, Britain and Germany between the World Wars* (Ithaca, NY: Cornell UP, 1984), and "Nationalism, the Mass Army, and Military Power," *International Security* 18, no. 2 (1993).

50. Katzenstein, *Culture of National Security*; Alexander Wendt, *Social Theory of International Politics* (Cambridge: Cambridge UP, 1999).

51. John Glenn, Darryl Howlett, and Stuart Poor, *Neorealism versus Strategic Culture* (London: Ashgate, 2004); John S. Duffield, Theo Farrell, Richard Price, and Michael C. Desch, "Isms and Schisms: Culturalism versus Realism in Security Studies," *International Security* 24, no. 1 (Summer 1999), pp. 156–80.

52. Farrell, "Culture and Military Power," pp. 24, 409; Jeannie L. Johnson, *Strategic Culture: Refining the Theoretical Construct* (Defense Threat Reduction Agency, SAIC, 2006), pp. 3, 9, 11.

53. Colin Dueck, "Realism, Culture and Grand Strategy: Explaining America's Peculiar Path to World Power," *Security Studies* 14, no. 2 (April–June 2005), pp. 195–231; Gideon Rose, "Neoclassical Realism and Theories of Foreign Policy," *World Politics* 51, no. 1 (October 1998), pp. 144–72; Michael Doyle, *Ways of War and Peace* (New York: W. W. Norton, 1997).

54. Jeffrey S. Lantis, *Strategic Culture: From Clausewitz to Constructivism* (Defense Threat Reduction Agency, SAIC, 2006).

55. Johnson, *Strategic Culture*, pp. 5–11.

56. Ibid., pp. 14–15, 20–21.

57. See Stephen Biddle, "Speed Kills? Reassessing the Role of Speed, Precision and Situational Awareness in the Fall of Saddam," *Journal of Strategic Studies* 30, no. 1 (2007), pp. 3–46, and *The 2006 Lebanon Campaign and the Future of Warfare* (Carlisle, PA: US Army War College, Strategic Studies Institute, 2008); special issue of the *Journal of Strategic Studies* 3 (June 2005).

58. Saul Bronfeld, "Did TRADOC Outmaneuver the Maneuverists? A Comment," *War and Society* 27, no. 2 (October 2008), pp. 111–25.

59. For Russian perspectives, see A. A. Kokoshin, *Formirovanie politiki assimetrichnogo otveta na SOI* (Moscow: URSS, 2008); Oleg Grinevskii, *Perelom: ot Brezhneva k Gorbachevu* (Moscow: Olma Press, 2004); Anton Pervushyn, *Zvezdnie voiny: Amerikanskaia respublika protiv Sovetskoi imperii* (Moscow: Exmo, 2005), and "Zvezdnie voiny: illiuzii i opasnosti," *Voennaia Mysl'* 9 (1985), pp. 15–28. For US reflections, see Mahn-

ken, *Technology and the American Way of War*, pp. 15–60 and 122–57; Leopoldo Nuti, *The Crisis of Détente in Europe* (New York: Routledge, 2008), pp. 86–124.

Chapter 1

1. Bernard Brodie, *War and Politics* (New York: Macmillan, 1973), p. 332.

2. This led to the notion of the archetypal personality. Georg Wilhelm Friedrich Hegel, *Phenomenology of Spirit* (New York: Oxford UP, 1977); Wilhelm Wundt, *Elements of Folk Psychology* (New York: Macmillan, 1921); Margaret Mead and Rhoda Métraux, *The Study of Culture at a Distance* (Chicago: U of Chicago P, 1953); William Sumner, *Folkways* (New York: New American Library, 1960); Albert Bandura, *Social Foundations of Thought and Action* (Englewood Cliffs, NJ: Prentice Hall, 1986); Edward T. Hall, *Beyond Culture* (New York: Anchor Press, 1976).

3. Hall, *Beyond Culture*; and *The Silent Language* (New York: Anchor Press, 1973), and *The Dance of Life: The Other Dimension of Time* (New York: Anchor Press, 1983).

4. Cotte Ratneshwa, "Juggling and Hopping: What Does It Mean to Work Polychronically?"; Allen Bluedorn, Thomas Kalliath, Michael Strube, and Gregg Martin, "Polychronicity and the Inventory of Polychronic Values (IPV)"; Mary Waller, Robert Giambatista, and Mary Zellmer-Bruhn, "The Effects of Individual Time Urgency on Group Polychronicity"; Charles Benabou, "Polychronicity and Temporal Dimensions of Work in Learning Organizations"; Carol Kaufman-Scarborough and Jay Lindquist, "Time Management and Polychronicity," all in *Journal of Managerial Psychology* 14, no. 3/4 (1999); Joseph E. McGrath and Nancy L. Rotchford, "Time and Behavior in Organizations," *Research in Organizational Behavior* 5 (1983), pp. 57–101.

5. Charles Cogan, *French Negotiating Behavior: Dealing with* La Grande Nation (Washington, DC: United States Institute of Peace, 2003); Richard Solomon, *Chinese Negotiating Behavior: Pursuing Interest through "Old Friends"* (Washington, DC: United States Institute of Peace, 1999); Robert Bathurst, *Intelligence and the Mirror: On Creating the Enemy* (New York: Sage, 1993); Raymond Cohen, *Negotiating across Cultures: International Communication in an Independent World* (Washington, DC: United States Institute of Peace Press, 1997); Edmund Glenn and Christine Glenn, *Man and Mankind: Conflict and Communication between Cultures* (Norwood, NJ: Ablex, 1981).

6. Geert Hofstede, *Culture's Consequences: International Differences in Work-Related Values* (Beverly Hills, CA: Sage, 1980), *Cultures and Organizations: Software of the Mind* (New York: McGraw-Hill, 2004), and *Culture's Consequences: Comparing Values, Behaviors, Institutions, and Organizations across Nations* (Thousand Oaks, CA: Sage, 2006). I have chosen only the relevant cultural parameters from Hofstede. For the full list of cultural dimensions offered by Hofstede, see all the above-mentioned sources and Hofstede's personal Web site: http://www.geert-hofstede.com/.

7. Aleksandr Luria, *Cognitive Development: Its Cultural and Social Foundations* (Cambridge, MA: Harvard UP, 1976); Zakharia Dershowitz, "Sub-cultural Patterns

and Psychological Differentiation," *International Journal of Psychology* 6 (1971), pp. 223–31; Johan Galtung, "Structure, Culture and Intellectual Style," *Social Science Information* 20 (1981), pp. 817–56; Lawrence Hirschfeld, *Race in the Making: Cognition, Culture, and the Child's Construction of Human Kinds* (Cambridge, MA: MIT Press, 1996); Joan Miller, "Culture and the Development of Everyday Social Explanation," *Journal of Personality and Social Psychology* 46 (1984), pp. 961–78; Bradd Shore, *Culture in Mind: Cognition, Culture, and the Problem of Meaning* (New York: Oxford UP, 1996).

8. Kenneth Goldstein and Sheldon Blackman, *Cognitive Style: Five Approaches and Relevant Research* (New York: Wiley, 1978); Richard Riding and Stephan Rayner, *Cognitive Styles and Learning Strategies* (London: David Fulton, 1998).

9. Steven Sloman, "Feature-Based Induction," *Cognitive Psychology* 25 (1993), pp. 231–80; "The Empirical Case for Two Systems of Reasoning," *Psychological Bulletin* 119 (1996), pp. 30–52; Richard Nisbett, Kaiping Peng, Incheol Choi, and Ara Norenzayan, "Culture and Systems of Thought: Holistic vs. Analytic Cognition," *Psychological Review* 108 (2001), pp. 291–310; Herman Witkin, "Field Dependence and Interpersonal Behavior," *Psychological Bulletin* 84 (1977), pp. 661–89.

10. Psychology provides the following explanation: individuals from high-context cultures live in a complex, interdependent social world with prescribed roles for group-centered relations. Attention to context is important to effective functioning. In contrast, participants from more independent and autonomous low-context cultures live in less constraining social worlds and are inclined to attend to the object and to define their goals with respect to it. The causal mechanism, according to cultural psychologists, operates as follows: social structure directs a person's attention to some aspects of the field at the expense of others. What is attended to influences metaphysics, that is, beliefs about the nature of the world and about causality. Metaphysics guides tacit epistemology, that is, beliefs about what it is important to know and how knowledge can be obtained. Epistemology dictates the development and application of some cognitive processes at the expense of others. Nisbett et al., "Culture and Systems of Thought"; Richard Nisbett and Takahiko Masuda, "Culture and Point of View," *Proceedings of the National Academy of Sciences* 100, no. 19 (2003), pp. 11163–11170; Richard Nisbett and Yuri Miyamoto, "The Influence of Culture: Holistic vs. Analytical Perception," *Trends in Cognitive Sciences* 9, no. 10 (October 2005); Ronald Rensink, "To See or Not to See: The Need for Attention to Perceive Changes in Scenes," *Psychology Science* 8 (1997), pp. 368–73; Harry Triandis, "The Self and Social Behavior in Different Cultural Contexts," *Psychology Review* 96 (1997), pp. 269–89.

11. *Gemeinschaft* (collectivist pattern) and *Gesellschaft* (individualist pattern) are sociological categories introduced by the German sociologist Ferdinand Tönnies for two types of human association. Gemeinschaft is a social association in which individuals are oriented to the larger association as much as if not more than to their own self-interest. Individuals in Gemeinschaft are regulated by common norms, or beliefs

about the appropriate behavior and responsibility of members of the association, to each other and to the association at large. Associations are marked by the "unity of will." Gesellschaft, in contrast, describes social associations in which, for the individual, the larger community never takes on more importance than individual self-interest, and lacks Gemeinschaft's level of shared mores. Gesellschaft is maintained through individuals acting in their own self-interest. Ferdinand Tönnies, *Community and Civil Society*, ed. Jose Harris (Cambridge: Cambridge UP, 2001). People with holistic styles of thinking come from backgrounds in which neither equality among persons nor differentiation of roles is as accentuated as they are in the background of those with analytical patterns of thinking. Edward Stewart and Milton Bennett, *American Cultural Patterns: A Cross-Cultural Perspective* (Yarmouth, ME: Intercultural Press, 1991), pp. 42–43.

12. Richard Nisbett, "Culture and Cognition," in Hal Pashler, ed., *Stevens' Handbook of Experimental Psychology*, vol. 2, *Cognition* (New York: John Wiley & Sons, 2002), pp. 561–97; Eduardo Wilner, "Comparing Traditions: Parallels between Eastern Mysticism and Western Science," *Comparative Civilizations Review*, no. 55 (Fall 2006), pp. 18–36. The essence of RMA—to highlight revolutionary versus evolutionary change in military affairs—is more dialectical than logical-analytical.

13. Edward Stewart and Milton Bennett, *American Cultural Patterns: A Cross-Cultural Perspective* (Yarmouth, ME: Intercultural Press, 1991), pp. 43–44.

14. Rosalie Cohen, "Conceptual Styles, Culture Conflict and Nonverbal Tests of Intelligence," *American Anthropologist* 71, no. 5 (October 1969), pp. 828–56.

15. Russian thinking style was found to be closer to the logical-analytical end of the continuum, when compared to the Chinese cultural mind-set, which manifests a stronger inclination toward a holistic-dialectical way of thought. Snejina Michailova and Verner Worm, "Personal Networking in Russia and China: Blat and Guanxi," *European Management Journal* 21, no. 4 (2003), pp. 509–19.

16. Gerhard Maletzke, *Interkulturelle Kommunikation: Zur Interaktion zwischen Menschen verschiedener Kulturen* (Opladen: Westdeutscher Verlag, 1996), cited in Stephan Dahl, "Intercultural Research: The Current State of Knowledge," Middlesex University Discussion Paper no. 26 (2004).

17. Loren Graham and Jean-Michel Kantor provided a good example for this argument comparing two different cultural approaches to mathematics, particularly the works on the descriptive theory of sets. Unlike their European colleagues, who subscribed to a rational, secular worldview and consequently did not attend to infinite numbers as legitimate mathematical objects, Russians took a different approach to the same problem. Not without considerable inclination toward an irrational mystical worldview, Russian mathematicians sought to link and to integrate logical, philosophical, theological, and psychological components to produce an innovative mathematical result. In the end, the Moscow school of mathematics gave birth to a new

field, descriptive set theory. Loren Graham and Jean-Michel Kantor, "A Comparison of Two Cultural Approaches to Mathematics, France and Russia," *ISIS—The History of Science Society* 97, no. 1 (March 2006), pp. 56–74, and "Soft Area Studies versus Hard Social Science: A False Opposition," *Slavic Review* 66, no. 1 (Spring 2007), pp. 1–20. For rationality and mysticism in the Russian way of thinking, see Bernice Rosenthal, *The Occult in Russian and Soviet Culture* (Ithaca, NY: Cornell UP, 1997), p. 155.

18. Thomas Mahnken, "Uncovering Foreign Military Innovation," *Journal of Strategic Studies* 22, no. 4 (December 1999), pp. 30–31; realizing that revolutionary discontinuity versus incremental evolution is under way may also originate in the subsequent two stages.

19. Richard Hundley, *Past Revolutions, Future Transformation: What Can the History of Revolutions in Military Affairs Tell Us about Transforming the US Military?* (Santa Monica, CA: RAND, 1999), p. 21.

20. Max Boot, *War Made New: Technology, Warfare and the Course of History, 1500 to Today* (New York: Gotham Books, 2006), p. 10.

21. Hundley, *Past Revolutions, Future Transformation*, p. 9.

22. Thomas Kuhn, *The Structure of Scientific Revolutions* (Chicago: U of Chicago P, 1996); Gregory Bason, *Steps to an Ecology of Mind* (Chicago: U of Chicago P, 2000), pp. 139–41.

23. Hundley, *Past Revolutions, Future Transformation*, p. 9.

24. Emily O. Goldman, "Introduction: Military Diffusion and Transformation," in Emily Goldman and Thomas Mahnken, *The Information Revolution in Military Affairs in Asia* (New York: Palgrave Macmillan, 2004), p. 7.

25. Theo Farrell, *The Sources of Military Change: Culture, Politics, Technology* (London: Lynne Rienner Publishers, 2002), p. 6.

26. Emily Goldman and Leslie Eliason, eds. *Diffusion of Military Technology and Ideas* (Stanford, CA: Stanford UP, 2003).

27. Hundley, *Past Revolutions, Future Transformation*, p. 25.

28. Robert K. Merton and Elinor Barber, *Travels and Adventures of Serendipity: A Study of Sociological Semantics and the Sociology of Science* (Princeton, NJ: Princeton UP, 2004); Robert Roystons, *Serendipity: Accidental Discoveries in Science* (New York: Wiley, 1999).

29. Elena Bondareva, *Green Building in the Russian Context: An Investigation into the Establishment of a Building System in the Russian Federation* (master's thesis, Cornell University, 2005), pp. 55–56.

Chapter 2

1. This chapter is based on Dima Adamsky, "Through the Looking Glass: The Soviet Military-Technical Revolution and the American Revolution in Military Affairs," *Journal of Strategic Studies* 31, no. 2 (April 2008). http://www.informaworld.com/

smpp/content~db=all~content=a791764957. William E. Odom, *The Collapse of the Soviet Military* (New Haven: Yale UP, 1998), pp. 72–75, and "Soviet Force Posture: Dilemmas and Directions," *Problems of Communism* 34, no. 4 (June–August 1985), pp. 1–14. The West learned a great deal about this Soviet doctrine from the translation of Marshal Sokolovskii's *Military Strategy*. Harriet Fast Scott, ed., *Soviet Military Strategy* (New York: Crane, Russak, 1975); V. D. Sokolovskii, *Voennaia strategiia* (Moscow: Voenizdat, 1962).

2. Donn A. Starry, "Extending the Battlefield," *Military Review* 61, no. 3 (1981), pp. 31–50, and "To Change the Army," *Military Review* 63, no. 3 (1983); Department of the Army, *Field Manual 100-5 Operations* (Washington, DC: GPO, 1986); Richard Lock-Pullan, "How to Rethink War: Conceptual Innovation and AirLand Battle Doctrine," *Journal of Strategic Studies* 4 (2005), pp. 679–702; Odom, *Collapse of the Soviet Military*, pp. 74–76.

3. Department of the Army, *Field Manual 100-5 Operations*, pp. i–1; Starry, "Extending the Battlefield," p. 46; Kimberly Marten Zisk, *Engaging the Enemy: Organization Theory and Soviet Military Innovation, 1955–1991* (Princeton, NJ: Princeton UP), pp. 121–32; John Romjue, *From Active Defense to AirLand Battle: The Development of Army Doctrine 1973–1982* (Fort Monroe, VA: US Army TRADOC, 1984), pp. 24–25.

4. Zisk, *Engaging the Enemy*, p. 142; Bernard W. Rogers, "Follow-On Forces Attack: Myths and Realities," *NATO Review* 32, no. 6 (1984), pp. 1–9; Lock-Pullan, "How to Rethink War," pp. 687–88.

5. Odom, *Collapse of the Soviet Military*, p. 76; Zisk, *Engaging the Enemy*, p. 134.

6. V. Kozhin and V. Trusin, "Voprosy primeneniia vooruzhennykh sil v operatsiiakh," *Zarubezhnoe voennoe obozrenie* (hereafter, *ZVO*), no. 10 (1983), pp. 18–19.

7. Christian Nunlist, *Cold War Generals: The Warsaw Pact Committee of Defense Ministers, 1969–90* (n.p.: Parallel History Project on NATO and the Warsaw Pact, 2001), pp. 14–15.

8. N. G. Popov, "Dostizhenie zhivuchesti voisk v operatsiiakh," *Voennaia Mysl'* (hereafter, *VM*), no. 1 (1983), pp. 32–44; S. Yegorov, "Mekhanizirovannaia diviziia SShA v nastuplenii," *ZVO*, no. 4 (1984), pp. 23–28; V. Lamkhin, "Vozdushnaia nastupatel'naia operatsiia," *ZVO*, no. 11 (1984), pp. 47–54; V. Sidorov, "Vedenie operatsii c primeneniem obychnykh sredstv porazheniia," *ZVO*, no. 1 (1985), pp. 7–15.

9. MacGregor Knox and Williamson Murray, *The Dynamics of Military Revolution, 1300–2050* (Cambridge: Cambridge UP, 2001), pp. 1–2.

10. V. A. Zolotarev, *Istoriia voennoi strategii Rossii* (Moscow: Institut Voennoi Istorii MO RF, 2000), p. 380.

11. As the General Staff war game *Zapad 77* indicated, the Soviets did not figure new Western doctrinal postulates and weaponry associated with the Active Defense doctrine into their war-gaming until at least 1977; see *Materialy razbora operativno-stratgecheskogo*

komandno shtabnogo uchenia 'Zapad-77' (Moscow: Ministerstvo Oborony SSSR, 1977), pp. 10–11.

12. V. F. Krest'ianinov, *Nauchno tekhnicheskaia revoliutcia i revoliutcia v voennom dele* (Leningrad: Voennaia Akademiia Tyla i Transporta, 1971), pp. 14–31; I. I. Anureev, *Nauchno tekhnicheskii progress i ego ispolzovanie v voennom dele* (Moscow: Obshestvo Znanie, 1982); A. Z. Gilmanov, *O Nauchno tekhnicheskoi revoliutcii kak vazhnom faktore sozdaniia materialn-tekhnicheskoi bazy komunizma* (Kazan': Kazanskoe Vysshee Voennoe Inzhenernoe Uchilishe, 1982); V. M. Bondarenko, *Nauchno tekhnicheskii progress i voennoe delo* (Moscow: Centralnyi Dom SA im. Frunze, 1973), pp. 26–41.

13. G. P. Otuytsskiy, "K Voprosu o sushchnosti voenno-tekhnicheskikh revoliutsii," *VM*, no. 2 (1998); Stephen Peter Rosen, *Winning the Next War: Innovation and the Modern Military* (Ithaca, NY: Cornell UP, 1994), pp. 257–58; P. M. Derevianko, *Revoliutsiia v voennom dele: vchem ee sushchnost'?* (Moscow: Ministerstvo Oborony SSSR, 1967); F. F. Gaivoronovskii, *Evoliutsiia voennogo iskusstva: etapy, tendentsii, printsipy* (Moscow: Voenizdat, 1987).

14. A. O. Baranov, *Sovremennaia voenno-tekhnicheskaia revoliutcia, ee soderzhanie i osobennsoti* (PhD diss., Voenno Politicheskaia Akademiia, Moscow, 1974), and *Sushnost' sovremennoi voenno-tekhnicjheskoi revoliutzii* (Moscow: Ministerstvo Oborony, 1977), pp. 16–18, 28–32; Iu. V. Man'ko, *Dialektika razvitiia sposobov i form vooruzhennoi bor'by* (Leningrad: Voennaia Akademia Sviazi, 1975); Dimitr Trendafilov, *Ideologicheskie aspekty sovremennoi voenno-tekhnicheskoi revoliutcii* (PhD diss., Voenno-Politicheskaia Akademiia, Moscow, 1977); D. D. Gorbatenko, *Faktor vremeni v sovremennom boiu* (Moscow: Ministerstvo Oborony, 1972).

15. V. M. Bondarenko, *Sovremennaya nauka i razvitie voennogo dela* (Moscow: Voenizdat, 1976), pp. 94–95, 109; N. Ogarkov, "Sovetskaia voennaia nauka," *Krasnaia zvezda* (hereafter, *KZ*), 18 February 1978; N. V. Mikhalkin, *Logiko-gnoseologicheskii analiz voenno-tekhnicheskogo poznaniia* (PhD diss., Voenno-Politicheskaia Akademiia, Moscow, 1983).

16. Odom, "Soviet Force Posture," p. 7.

17. N. V. Ogarkov, *Vsegda v gotovnosti k zashchite otechestva* (Moscow: Voennoe Izdatel'stvo, 1982), pp. 31–43, 59–67, *Istoriia uchit bditel'nosti* (Moscow: Voenizdat, 1985), "Na strazhe mirnogo truda," *Kommunist*, no. 10 (1981), pp. 80–91; "Nadezhnaia zaschita mira," *KZ*, 23 September 1983, and "Pobeda i segodnaishnii den'," *KZ*, 9 May 1983; interview with Marshal N. Ogarkov, *KZ*, 9 May 1984; G. P. Otiutzkii, *Voenno tekhnicheskaia politika gosudarstva kak factor razvitiia sistemy 'chelovek-voennaia tekhnika* (Moscow: Voenno-Politicheskaia Akademiia, 1982).

18. M. A. Gareev, *M. V. Frunze—Voennyi teoretik* (Moscow: Voenizdat, 1985), pp. 425, 438–39, "Razvitiie form i metodov operativnoi i boevoi podgotovki v Sovetskoi Armii," *VM*, no. 2 (1987), pp. 38–50, and "Tvorcheskii kharakter sovestkogo voennogo isskustva v Velikuiu Otechestvennuiu Voinu," *Voyenno-istoricheskiy zhurnal* (hereaf-

ter, *VIZh*), no. 7 (July 1985), pp. 28–29. V. G. Reznichenko, *Taktika* (Moscow: Voenizdat, 1984), pp. 23–24; and "Vysokotochnoe oruzhie," in S. F. Akhromeev, ed., *Voennyi entsiklopedicheskii slovar* (Moscow: Voenizdat, 1986), p. 172.

19. M. G. Popkov, *Metodologicheskii analiz informacionnykh processov v sisteme chelovek-voennaia tekhnika* (PhD diss., Voenno-Politicheskaia Akademiia, Moscow, 1983); N. Nechaev, "Voennye sistemy sviazi: tendencii ikh razvitiia," *Tekhnika i vooruzheniia*, no. 6 (July 1986); Stanislaw Koziej, "Przewidywane kierunki zmian w taktyce Wojsk Ladowych," *Przeglad Wojsk Ladowych* (September 1986), pp. 1–8; Gen. Ye. Kolibernov, interview in *KZ*, 21 November 1985.

20. V. A. Gorbunov, *Effektivnost' obnaruzheniia celei* (Moscow: Ministerstvo Oborony, 1980), chs. 2 and 5; I. Vorob'yev, "Sovremennie vooruzheniia I taktika," *KZ*, 15 September 1984; and "Novie vooruzheniia i principi taktiki," *Sovetskoe voennoe obozrenie*, no. 2 (February 1987); V. Makarebskiy, "Blitzkrig v epokhu nauchno tekhnicheskoi revoliutcii," *Voennye Znaniia* (hereafter, *VZ*) no. 9 (1986); A. Dvoretskiy and V. Potashev, "O kontceptcii vozdushno-nazemnoi operatcii," *Vestnik PVO*, no. 8 (August 1984); V. P. Shipovalov, "Bor'ba s tankami," *VIZh*, no. 9 (September 1986).

21. V. Makarevskiy, "Gonka obychnyh vooruzhenii i problemi ee ogranicheniia," *Mirovaya ekonomika i mzhdunarodnye otnosheniia*, no. 5 (May 1984); I. Golushko, "Tyl v usloviah ispolzovania protivnikom vysoko-tochnogo oruzhiia," *Tyl i snabzhenie*, no. 7 (July 1984).

22. Reznichenko, *Taktika*; P. K. Altukhov, *Osnovy teorii upravleniya voyskami* (Moscow: Voenizdat, 1984), pp. 32–34.

23. Vojtech Mastny and Malcolm Byrne, *A Cardboard Castle? An Inside History of the Warsaw Pact, 1955–1991* (New York: CEU Press, 2005), p. 45.

24. Zolotarev, *Istoriia voennoi strategii Rossii*, p. 442; N. V. Ogarkov, "Voennaia strategiia," *Sovetskaia voennaia entsiklopediia* (Moscow: Voenizdat, 1979), vol. 7, pp. 564–65.

25. Oleg Grinevskii, *Stsenarii dlia tret'ei mirovoi* (Moscow: Olma Press, 2002), pp. 353–57; Georgii Kornienko, *Kholodnaia voina: svidetel'stvo ee uchastnika* (Moscow: Olma Press, 2001), pp. 363–67; N. V. Ogarkov, *Pravda*, 6 December 1983; Zolotarev, *Istoriia voennoi strategii Rossii*, p. 446; Makhmut Gareev, *Esli zavtra voina* (Moscow: Vladar, 1995), pp. 86–88; Gareev, *M. V. Frunze—Voennyi teoretik*, pp. 239–41; Ogarkov interview, *KZ*, 9 May 1984.

26. I. Vorob'iev, "Sovremennye vooruzheniia i taktika," *KZ*, 15 September 1984; V. M. Bondarenko, "Nauchno-tekhnicheskii progress i voennoe delo," *Kommunist vooruzhennykh sil* (November 1986), pp. 13–21; S. F. Akhromeev, "Prevoskhodstvo sovetskoi voennoi nauki i sovetskogo voennogo iskusstva—odin iz naibolee vazhnykh faktorov pobedy," *Kommunist* (February 1985).

27. Bondarenko, "Nauchno-tekhnicheskii progress i voennoe delo," p. 14; Ogarkov, interview, *KZ*, 1984; Iu. Molostov, "Vysokotochnoe protivotankovoe oruzhie i

obshevoiskovoi boi," *Voennyi vestnik* (hereafter, *VV*) (October 1986); Reznichenko, *Taktika*, p. 128; Akhromeev, "Prevoskhodstvo sovetskoi voennoi nauki i sovetskogo voennogo iskusstva"; Gareev, *M. V. Frunze—Voennyi teoretik*, pp. 239–40.

28. N. V. Ogarkov, "Zashchita sotsializma: istoricheskii opyt i tekushchii moment," *KZ*, 9 May 1984, *Istoriia uchit*, pp. 68–69, and *Vsegda v gotovnosti*, p. 16.

29. Odom, *Collapse of the Soviet Military*, p. 69; Mastny and Byrne, *Cardboard Castle*, p. xl.

30. Mastny and Byrne, *Cardboard Castle*, p. xliii; Zolotarev, *Istoriia voennoi strategii Rossii*, pp. 407–19, 456; Gareev, *Esli zavtra voina*, pp. 80–81. The scenario of the conventional confrontation was created under the assumption that war may still become nuclear. Zolotarev, *Istoriia voennoi strategii Rossii*, pp. 407–19, 456; V. I. Varennikov, *Nepovtorimoe*, (Moscow: Art Press, 2001), vol. 4, ch. 3; Gareev, *Esli zavtra voina*, pp. 80–81; L. I. Ol'shtynskii, *Vzaimodeistvie armii i flota* (Moscow: Voenizdat, 1983), pp. 6–8; V. A. Merimskii, *Takticheskaia podgotovka motorstelkovykh i tankovykh podrazdelenii* (Moscow: Voenizdat, 1984), pp. 8–9. Moreover, the belief in survival in a nuclear war and in some form of victory was expressed by the Soviet military even in the early 1980s. A lack of faith would mean pacifism, defeatism, and lassitude of the Soviet military effort. Such behavior the Soviet military could not abide, even if it sincerely considered such a belief to be completely irrational. Grinevskii, *Stsenarii*, pp. 353–57; Fritz Ermarth, "Contrasts in American and Soviet Strategic Thought," *International Security* 3, no. 2 (Autumn 1978), pp. 138–55.

31. Zolotarev, *Istoriia voennoi strategii Rossii*, pp. 379–80; *Programma Kommunisticheskoi Partii SSSR* (Moscow: Politizdat, 1975), p. 97. Conventional thinking started to emerge after the Cuban Missile Crisis. Warsaw Pact forces chief of staff Vitalii Gribkov, cited in Odom, *Collapse of the Soviet Military*, pp. 65–69, 433.

32. A. G. Khar'kov, "K voprosu o nachal'nom periode voiny," *VM*, no. 8 (1984), pp. 25–34; M. M. Kozlov, "Organizatsiia i vedenie strategicheskoi oborony po opytu Velikoi Otechestvennoi Voiny," *VIZh*, no. 12 (1980), pp. 9–17, and *Akademiia General'nogo Shtaba* (Moscow: Voenizdat, 1987), pp. 184–87.

33. V. V. Turchenko, "Tendentsii razvitiia teorii i praktiki strategicheskoi oborony," *VM*, no. 8 (1979), pp. 13–24; "O strategicheskoi oborone," *VM*, no. 7 (1982), pp. 16–27; and also see in I Vorob'ev, "Sootnoshenie i vzaimosviaz' nastuplenia i oborony," VM, no.4 (1980), pp. 53–55; N. V. Ogarkov, "Pobeda i segodniashnii den'," *Izvestiia*, 9 May 1983; Zolotarev, *Istoriia voennoi strategii Rossii*, pp. 409–10, 457; Koziej, "Przewidywane kierunki zmian," pp. 1–8.

34. A. Z. Ekimovskii, "Puti sozdaniia ustoichivoi i aktivnoi oborony," *VM*, no. 7 (1983), pp. 19–28; Koziej, "Przewidywane kierunki zmian," p. 4; I. N. Manzhurin, "Otrazhenie kontrudarov protivnika v khode nastupatel'noi operatsii," *VM*, no. 10 (1986), pp. 14–22; V. A. Nazarenko, "Narushenie upravleniia voiskami—vazhenaia boevaja zadacha," *VM*, no. 7 (1983), pp. 46–51.

35. I. Vorob'ev, "Sootnoshenie i vzaimosviaz' nastuplenia i oborony," *VM*, no. 4 (1980), pp. 49–59.

36. See the next section of this chapter.

37. I. G. Zavialov, "O roli boevogo potentsiala vooruzhennykh sil v nastuplenii i oborone," *VM*, no. 3 (1983), pp. 3–19; V. Matsulenko, "Nekotorye vyvody iz opyta nachial'nogo perioda Velikoi Otechestvennoi Voiny," *VIZh*, no. 3 (1984), pp. 35–43; M. A. Gareev, "Tvorcheskii kharakter sovetskoi voennoi nauki v Velikoi Otechestvennoi Voine," *VIZh*, no. 2 (1989), pp. 22–30; Kozlov, "Organizatsiia i vedenie strategicheskoi oboroni," pp. 9–17; P. Mel'nikov, "Operativnaia Maskirovka," *VIZh*, no. 4 (1982), pp. 18–26; A. Krupchenko, "Tekhnicheskoe obespechenie tankovykh korpusov, deistvovav-shikh v kachestve podvizhnikh grupp," *VIZh*, no. 6 (1982), pp. 18–26; A. P. Maryshev, "Nekotorie voprosy strategicheskoi oborony v Velikoi Otechestvennoi Voine," *VIZh*, no. 6 (1986), pp. 9–16.

38. I. G. Zavialov, "Oboronitel'naia napravlennost' Sovetskoi voennoi doktriny," *VM*, no. 1 (1981), pp. 15–26; A. A. Danilevich, "Voenno teoriticheskoe nasledie M. V. Frunze i sovremmennost," *VIZh*, no. 6 (1985), pp. 80–87; M. A. Gareev, "Ob opyte boevoi podgotovki voisk," *VIZh*, no. 4 (1983), pp. 11–20; see also Gareev, *M. V. Frunze—Voennyi teoretik*, pp. 214, 241–43, 437–38.

39. Zolotarev, *Istoriia voennoi strategii Rossii*, pp. 425, 474–76.

40. Khar'kov, "K voprosu o nachal'nom periode voiny," pp. 25–34; Kozlov, "Orga-nizatsiia i vedenie," pp. 9–17, and *Akademiia General'nogo Shtaba*, pp. 184–87.

41. Richard H. Van Atta, *Transformation and Transition: DARPA's Role in Foster-ing an Emerging Revolution in Military Affairs* (Alexandria, VA: Institute for Defense Analysis, 2003), pp. 16–18.

42. Mastny and Byrne, *Cardboard Castle*, pp. 46–47. Also see Thomas Mahnken, *Technology and the American Way of War* (New York: Columbia UP, 2008), pp. 124–25.

43. V. Filippov, "Soedineniia novogo tipa v armii SShA," *ZVO*, no. 7 (1978), p. 29; A. Solov'ev and L. Guliaev, "Radio-elektronnaia razvedka SShA," *ZVO*, no. 7 (1978), pp. 12–18; V. Afinov, "Sredstva REB sukhoputnykh voisk SShA," *ZVO*, no. 4 (1980), pp. 55–57.

44. A. Bulatov, "Bor'ba s tankami na bolshikh dal'nostiakh," *ZVO*, no. 12 (1979), pp. 12–13; N. Fomich, "Protivotankovye sredstva armii SShA," *ZVO*, no. 8 (1981), pp. 35–40.

45. V. Dmitriev and N. Germanov, "Upravliemye aviatsionnye bomby," *ZVO*, no. 3 (1981), pp. 55–60; B. Semenov, "Takticheskie upravliaemie rakety klassa vozdukh-poverkhnost," *ZVO*, no. 5 (1981), pp. 49–57.

46. I. Loshchilov, "Sredstva avtomatizatsii upravleniia voiskami v boiu," *ZVO*, no. 5 (1978), pp. 35–41; V. A. Aleksandrov, "O razvitii avtomatizirovannykh system upravle-niia v armii SshA," *VM*, no. 3 (1983), pp. 74–78.

47. V. Afinov, "Amerikanskaia sistema PLSS," *ZVO*, no. 4 (1980), pp. 55–57; V. Dmitriev, "Amerikanskaia sistema SOTAS," *ZVO*, no. 4 (1982), pp. 40–42; P. Isaev,

"Bor'ba s tankami," *ZVO*, no. 12 (1982), pp. 37–42; N. Dmitriev, "Vzaimodeistvie aviat-
sii s sukhoputnymy voiskami," *ZVO*, no. 6 (1980), p. 48; N. Stapenko, "Batal'ionnaia
takticheskaia gruppa v aktivnoi oborone," *ZVO*, no. 2 (1981), pp. 29–34; R. Simo-
nian, "Tendentsii v razvitii voennoi doktriny SShA," *ZVO*, no. 8 (1983), p. 15; G. Vasil'ev,
"Voprosy operativnogo iskusstva v vooruzhennykh silakh SShA," *ZVO*, no. 12 (1983),
pp. 3–7.

48. E. G. Evgeniev, "Novye napravleniia gonki vooruzhenii v stranah NATO,"
VM, no. 1 (1977), pp. 90–96; V. A. Tumas, "Voennoe iskusstvo sukhoputnykh voisk
NATO na sovremennom etape," *VM*, no. 8 (1977), pp. 79–87; K. M. Popov, "Zarubezh-
naia pechat' o kosmicheskom oruzhii," *VM*, no. 1 (1979), pp. 59–66; A. N. Ponomarev,
"Sostoianie i osnovnye napravleniia razvitiia aviatsionnoi tekhniki VVS SShA i dru-
gikh stran NATO," *VM*, no. 6 (1977), pp. 71–81; V. Borisov, "Organizatsiia voennykh
nauchno-tekhnicheskikh issledovanii v NATO," *VM*, no. 2 (1975), pp. 84–89; V. A.
Aleksandrov and V. A. Tumas, "Sovremennaia oborona po vzgliadam NATO," *VM*,
no. 10 (1978), pp. 89–96; M. V. Vasilchenko, "Operativnaia podgotovka vooruzhennykh sil NATO v 1980 godu," *VM*, no. 4 (1981), pp. 62–69; N. S. Nikolaev, "O kharaktere i soderzhanii operativnoi podgotovki ob'edinennykh sil NATO," *VM*, no. 6 (1980),
pp. 66–72.

49. Ogarkov interview, *KZ*, 9 May 1984.

50. Iu. I. Dmitriev and V. A. Mashchenko, "Priminenie navigatsionnikh sput-
nikovykh system SShA v voennykh tseliakh," *VM*, no. 10 (1983), pp. 79–80; A. A.
Zhovannik, "Kosmichiskie sistemy sviazi i ikh ispol'zovanie dlia upravleniia vooru-
zhennymy silami," *VM*, no. 4 (1983), pp. 34–42; N. I. Ivliev, "Operativnaia i boevaia
podgotovka vooruzhennykh sil NATO v 1982 godu," *VM*, no. 6 (1983), pp. 70–75; I. N.
Loschilov, "Perspektivy razvitiia ASU operativno takticheskogo naznacheniia sukhoputnykh voisk SShA," *VM*, no. 7 (1985), pp. 69–77, and "Amerikanskaia kontseptsiia
upravlenie, sviaz' i razvedka," *VM*, no. 7 (1986), pp. 63–72; V. Kozhin, "Voprosy primeneniia vooruzhennikh sil v operatsiiakh," *ZVO*, no. 10 (1983), pp. 18–19; N. Ivlev and
V. Viktorov, "Kompleksnoe uchenie voisk tsentral'noi gruppy armii NATO," *ZVO*, no.
9 (1983), p. 10; L. Levadov and V. Viktorov, "Manevry i ucheniia NATO—ugroza
miru," *ZVO*, no. 7 (1984), pp. 8–9, and "Itogi operativnoi podgotovki ob'edinnenykh
vooruzhennykh sil NATO v 1984 godu," *VM*, no. 3 (1985), pp. 64–72.

51. Golushko, "Tyl v usloviah ispolzovania protivnikom."

52. V. V. Afinov, "Razvitie v SShA vysokotochnogo oruzhiia i perspektivy sozda-
nia razvedovatel'no-udarnikh kompleksov," *VM*, no. 4 (1983), pp. 63–71; V. G. Krymt-
sev and Iu. I. Molostov, "Vysokotochnoe protivotankovoe oruzhie armii stran NATO
i perspektivy ego razvitiia," *VM*, no. 10 (1984), pp. 73–79; N. I. Ivliev and L. V. Levadov,
"Strategicheskoe KShU NATO Zima-83," *VM*, no. 12 (1983), pp. 70–73.

53. A. F. Volkov, "Leninskie printsypy voenno-tekhnicheskoi politiki KPSS," *VM*,
no. 4 (1980), pp. 31–38; I. Anureev, "Znachenie nauchno tekhnicheskoi revoliutsii dlia
ukreplenia oborony i povysheniia boevoi gotovnosti Sovetskikh Vooruzhennykh Sil,"

VM, no. 6 (1975), pp. 65–76, "O vzaimosviazi voennoi nauki s estevstvennymi naukami," *VM*, no. 1 (1981), pp. 27–35, "Vyshe effektivnost' i kachestvo voenno-nauchnykh znanii," *VM*, no. 5 (1983), pp. 3–16, and "KPSS i voennoe stroitel'stvo," *VM*, no. 9 (1983), pp. 73–78; M. M. Kir'ian, ed., *Voenno-tekhnicheskii progress i vooruzhennie sily SSSR* (Moscow: Voennoe Izdatel'stvo, 1982), pp. 262–64; I. E. Shavrov and M. I. Galkin, eds., *Metodologiia voenno-nauchnogo poznaniia* (Moscow: Voenizdat, 1977), pp. 3–5; Bondarenko, *Sovremennaia voennaia nauka*; V. V. Serebriannikov, *Osnovi Marksistsko-Leninskogo ucheniia o voine i mire* (Moscow: Voennoe Izdatel'stvo, 1982), chs. 2 and 4.

54. N. Maksimov, "Kontseptsiia 'vystrel-porazhenie," *ZVO*, no. 11 (1979), pp. 13–14; Y. G. Yevgen'iev, "Novye napravleniia gonki vooruzhenii v stranakh NATO," *VM*, no. 1 (1977), pp. 88–96; A. G. Sinitskii, "Nekotorie voprosy razvitiia vooruzheniia i boevoi tekhniki sukhoputnikh voisk NATO i ikh boevogo primeneniia," *VM*, no. 10 (1977), pp. 83–91; A. S. Baturin, "Nauchno tekhnicheskaia revoliutsiia i voennye prigotovleniia SShA," *VM*, no. 1 (1981), pp. 75–80.

55. Reznichenko, *Taktika*, ch. 2.

56. Ogarkov, *Vsegda v gotovnosti*, pp. 31–35, 40–43, 59–67, "Na strazhe mirnogo truda," pp. 80–91, "Nadezhnaia zaschita mira," *KZ*, 23 September 1983, and *Istoriia uchit bditel'nosti*.

57. Shavrov and Galkin, *Metodologiia*.

58. Shimon Naveh, *In Pursuit of Military Excellence: The Evolution of Operational Theory* (London: Frank Cass, 1997), pp. 166–67; L. I. Voloshin, "Teoriia glubokoi operatsii i tendentsii ee razvitiia," *VM*, no. 8 (1978), pp. 14–26; A. M. Maiorov, "Proriv oborony: teoriia i praktika mirovykh voin," *VM*, no. 5 (1978), pp. 79–94; R. Savushkin, "K voprosu o vozniknovenii i razvitii operatsii," *VIZh*, no. 5 (May 1979), and "K voprosu o zarozhdenii teorii posledovatel'nikh nastupatelnikh operatsii," *VIZh*, no. 5 (May 1983), pp. 12–20.

59. N. V. Ogarkov, "Glubokaia operatsiia (boi)," in *Sovetskaia voennaia entsiklopedia*, vol. 2, 1976, pp. 574–78.

60. A. F. Bulatov, "Aktual'nye voprosy sovremmennogo nastupatel'nogo boia," *VM*, no. 11 (1984), pp. 60–69; V. F. Mozolev, "Ob obshchikh osnovakh teorii sovetskogo operativnogo iskusstva," *VM*, no. 3 (1979), pp. 13–22; M. I. Bezkhrebetnii, "Sovmestnaia operatsiia—glavnaia forma sovremennikh boevikh deistvii," *VM*, no. 7 (1979), pp. 27–34, and "Podgotovka posleduiushchikh nastupatel'nikh operatsii," *VM*, no. 7 (1982), pp. 28–38.

61. Ogarkov, *Vsegda v gotovnosti*, pp. 34–36; Odom, "Soviet Force Posture," pp. 7–8.

62. Ogarkov, *Vsegda v gotovnosti*, pp. 44–45; S. Petrov, "Puti dostizhenia vysokikh tempov nastupleniia,"*VM*, no. 1 (1975); V. I. Varennikov, "Nekotorye problemy razvitiia uspekha v nastupatelnykh operaciiakh," *VM*, no. 8 (1979); I. D. Pombrik, "Aktualnye voprosy razvitia teorii nastupatel'noi operatsii," *VM*, no. 10 (1983); Gareev, *Esli zavtra voina*, p. 83.

63. A. G. Khar'kov, "Voevat' ne chislom a umeniem," *VM*, no. 6 (1983), pp. 34–42; P. G. Skachko, "Odnovremennoe vozdeistvie na vsiu glubinu operativnogo postroeniia protivnika—vedushchaia tendentsiia v razvitii teorii operativnogo iskusstva," *VM*, no. 7 (1985), pp. 18–24.

64. I. N. Vorob'ev, "Novoe oruzhie i razvitie printsipov obshchevoiskovgo boia," *VM*, no. 6 (1986), pp. 35–37; I. Vorob'ev, "Sovremennoe vooruzhenie i taktika," *KZ*, 15 September 1984; Gareev, *M. V. Frunze—Voennyi teoretik*, p. 245.

65. Reznichenko, *Taktika*, ch. 2; G. E. Peredel'skii, "Tendentsii razvitiia polevoi artilerii NATO," *VM*, no. 11 (1983), pp. 62–69; V. Shabanov, "Material'naia osnova oboronnoi moshchi," *KZ*, 15 August 1985; V. V. Tkachev, "Vzaimodeistvie v nastupatel'nom boiu," *VM*, no. 8 (1983), pp. 51–56; V. V. Krysanov, "Massirovanie sil i sredstv na glavnykh napravleniiakh-iskusstvo i raschet," *VM*, no. 5 (1984), pp. 26–32; M. Belov and V. Shchukin, "Razvedovatel'no-porazhayushchiye kompleksy armii SShA," *VV*, no. 1 (1985), pp. 86–89; Iu. Molostov, "Zashchita ot vysokotochnogo oruzhiia," *VV*, no. 2 (1987), pp. 83–86; Skachko, "Odnovremennoye vozdeystvie." For Western analyses of the Soviet ROK and RUK concept, see: Cohen, *Gulf War Air Power Survey Summary Report*, p. 237; Larry A. Brisky, "The Reconnaissance Destruction Complex: A Soviet Operational Response to Air-Land Battle," *Soviet Military Studies* 2 (1990), pp. 297–98; Naveh, *In Pursuit of Military Excellence*, pp. 166–67; Odom, "Soviet Force Posture," p. 9.

66. Zolotarev, *Istoriia voennoi strategii Rossii*, p. 475.

67. *Armeiskii Pregled*, no. 1 (1983), p. 98, and no. 2 (1983), p. 79.

68. A. Karemov, "Voennaia doktrina SShA," *ZVO*, no. 4 (1983), p. 11; Afinov, "Razvitie v SShA vysokotochnogo oruzhiia," pp. 69–74. The author noted the US ALB and ROK and RUK as the operational core of this modern doctrine. Skachko, "Odnovremennoye vozdeystvie," pp. 18–24; Ye. G. Korotchenko, "Ob evoliutsii printsipov voennogo iskusstva," *VM*, no. 9 (1988), pp. 22–23.

69. *Razvedovatel'no udarnyi kompleks*, the term in *Voennyi entsiklopedicheskii slovar'* (Moscow: Voenizdat, 1986).

70. Timothy L. Thomas, "Information Warfare in the Second Chechen War: Motivator for Military Reform?" in Anne C. Aldis and Roger N. McDermott, *Russian Military Reform, 1992–2002* (London: Frank Cass, 2003), pp. 216–17; Brisky, "Reconnaissance Destruction Complex."

71. Jeffrey McKitrick, "The Revolution in Military Affairs," in *Battlefield of the Future: 21st Century Warfare Issues* (Maxwell AFB, AL: Air UP, September 1995).

72. Jeffrey R. Cooper, "Another view of the Revolution in Military Affairs," in *In Athena's Camp: Preparing for Conflict in the Information age*, ed. John Arguilla and David Ronfeldt (Rand: National Defense Research Institute, 1997). p. 125; and see the following section of this chapter.

73. Robert Tomes, *Military Innovation and the Origins of the American Revolution in Military Affairs* (PhD diss., Department of Government and Politics, University of Maryland, 2004), pp. 160–63.

74. General Makhmut Gareev, then a head of the military science directorate, was one of the conceptual founding fathers of the OMG concept. The acronym *OMG* was even sometimes translated as "Gareev's Operational Thought" (*"operativnaia mysl' Gareeva"*). Interview with Makhmut Gareev, Moscow, July 2006. I. Krupchenko, "Kharakternye cherty razvitiia i primeneniia tankovykh voisk," *VIZh*, no. 9 (1979), pp. 25–32.

75. Interview with Gareev; Tkachev, "Vzaimodeistvie v nastupatel' nom" Krysanov, "Massirovanie sil i sredstv," pp. 26–32; Naveh, *In Pursuit of Military Excellence*, pp. 166–67; Odom, "Soviet Force Posture," p. 9.

76. Mastny and Byrne, *Cardboard Castle*, p. 482; interview with Gareev; A. Babadzhanian, *Tanki i tankovie voiska* (Moscow: Voenizdat, 1980), ch. 2; Reznichenko, *Taktika*, ch. 3, pp. 152–63.

77. Mastny and Byrne, *Cardboard Castle*, pp. xlii, 482.

78. Zygmunt Czarnotta, "Uzycie artylerii w wojnach lokalnych," *Przeglad Wojsk Ladowych*, vols. 5–6, 1987.

79. V. I. Varennikov, *Nepovtorimoe* (Moscow: Art Press, 2001), vol. 4.

80. Ibid., ch. 3.

81. Varennikov, *Nepovtorimoe*, vol. 4.

82. Vojtech Mastny, *Oral History Interviews with Polish Generals* (Washington, DC: PHP Publication Series, 2002).

83. V. V. Zhurkin, S. A. Kaganov, and A. V. Kortunov, "O razumnoi dostatochnosti," *SSha: Ekonomika, Politika, Ideologiia*, no. 12 (December 1987).

84. Ogarkov, *Istoriia uchit bditel'nosti*, pp. 49–50.

85. Stephen Blank, "Preconditions for a Russian RMA: Can Russia Make the Transition?" *National Security Studies Quarterly* 6, no. 2 (Spring 2000), pp. 1–28.

86. Interview with Gareev; M. A. Gareev, *Afganskaia strada* (Moscow: INSAN, 1999), chs. 1 and 5.

87. Lester Grau, trans. and ed., *The Bear Went over the Mountain: Soviet Combat Tactics in Afghanistan* (Washington, DC: Diane, 1996).

88. Robert Cassidy, *Russia in Afghanistan and Chechnya: Military Strategic Culture and the Paradoxes of Asymmetric Conflict* (Carlisle, PA: US Army War College, 2003).

89. Varennikov, *Nepovtorimoe*.

90. Fritz Gaenslen, "Culture and Decision Making in China, Japan, Russia, and the United States," *World Politics* 39 (1986), pp. 78–103; Oleg Kharkhordin, *The Collective and the Individual in Russia* (Berkeley: U of California P, 1999). For a different interpretation, which emphasizes the individual over the collective, see Laura Engelstein and Stephanie Sandler, *Self and Story in Russian History* (Ithaca, NY: Cornell UP, 2000).

91. Yale Richmond, *From Nyet to Da: Understanding the Russians* (Yarmouth, ME: Intercultural Press, 1996), pp. 5–26; Bondareva, *Green Building*, pp. 53–54; Christoph

Neidhart, *Russia's Carnival: The Smells, Sights, and Sources of Transition* (New York: Rowman and Littlefield, 2003), pp. 106–7, 128, 131, 173; Geert Hofstede, "Cultural Constraints in Management Theories," *Academy of Management Executive* 7, no. 1 (1993), pp. 81–94; Manfred F. R. Kets de Vries, I. Gurkov, and Y. Kuz'minov, "Organizational Learning in Russian Privatized Enterprises," *International Studies in Management and Organization* 25, no. 4 (1995), pp. 91–117; Alexei V. Matveev and Paul E. Nelson, "Cross Cultural Communication Competence and Multicultural Team Performance: Perceptions of American and Russian Managers," *International Journal of Cross Cultural Management* 8 (2004), pp. 253–70; Maureen Cote, *Russian Psychology in Transition: Interviews with Moscow Psychologists* (New York: Nova Science, 1998), pp. 51, 89, 106–9, 136–39, cited in Igor Grossmann, "Iron Curtain around the Russian Soul: Religion, Self-Reflection and Emotions," research paper, University of Michigan, 2006, pp. 7–8.

92. Betsy Stevens, "Russian Teaching Contracts: An Examination of Cultural Influence and Genre," *Journal of Business and Technical Communication* 4, no. 1 (January 2000), pp. 38–57; Patricia Hagen, "Teaching American Business Writing in Russia," *Journal of Business and Technical Communication* 12, no. 1 (1998), pp. 109–26; Frank T. Rothaermel, Suresh Kotha, and H. Kevin Steensma, "International Market Entry by U.S. Internet Firms," *Journal of Management* 32, no. 2 (2006), pp. 56–82; Maria Cseh, "Facilitating Learning in Multicultural Teams," *Advances in Developing Human Resources* 5, no. 2 (2003), pp. 26–40.

93. Robert Bathurst, *Intelligence and the Mirror: On Creating the Enemy* (New York: Sage, 1993); Neidhart, *Russia's Carnival*; Richmond, *From Nyet to Da*; T. Stefanenko, *Ethnopsikhologia* (Moscow: Aspekt Press, 2006); Grossmann, "Iron Curtain," p. 12.

94. Often-cited contradictions of Russian culture include: despotism, versus the hypertrophy of the state versus anarchism; a disposition toward violence, versus humanity and gentleness; a belief in ceremonies, versus a quest for truth; a heightened consciousness of personality, versus impersonal collectivism; a messianic search for God versus militant godlessness. On the contradictions of the Russian temperament, see: Bathurst, *Intelligence and the Mirror*, pp. 31–33; Grossmann, "Iron Curtain," pp. 7, 10; Cote, *Russian Psychology in Transition*, p. 84; V. F. Shapovalov, *Istoki i smysl rossiskoi civilizavii* (Moscow: Fair Press, 2003), pp. 210–12, 228–29; V. V. Rozanov, *Russkaia ideia* (Leningrad: RussPress, 1990), pp. 97–98; N. O. Losskii, *Kharakter russkogo naroda* (New York: Posev, 1957), pp. 38–40; I. A. Il'in, *Suschnost' i svoeobrazie russkoi kulturi* (Moskva: Art Press, 1996), p. 172; V. A. Zaicev, *Nacionalnoe svoeobrazii russkoi kulturi* (Kostroma: Kostromskoi Gosudarstvennyi Universitet, 2002), pp. 32–34, and "Universlanoe i specifichskoe v rossiiskoi imperii," *Obshetvennie nauki i sovremennost*, no. 3 (1993), pp. 87–88. Other scholars explain the inconsistency and complexity of the Russian character by pointing to the fact that in Russia two streams of world history—

East and West—merge and collide. Berdyaev, *Russkaia idea*, pp. 42–44; Gregory Dol-
gopolov, "World History in 60 Seconds: How Bank Imperial Re-Edited Russian Binary
Logic," *Social Semiotics* 10, no. 1 (2000), pp. 5–20.

95. Ermarth, *Russia's Strategic Culture*, p. 7: "From the old Soviet marshals to the
officer cadets, everyone argued that modern war was impossible without nuclear
weapons, that the Soviet Union would never start it, but on the other hand there was
no doubt and even operational plans existed on how to win the nuclear war that the
American imperialists would start." Bathurst, *Intelligence and the Mirror*, pp. 11, 87–89,
105–7.

96. G. Matthew Bonham, Victor Sergeev, and Pavel Parshin, "The Limited Test-
Ban Agreement: Emergence of New Knowledge Structures in International Negotia-
tions," *International Studies Quarterly* 41 (June 1997), pp. 215–40; Incheol Choi, Rich-
ard Nisbett, Ara Norenzayan, and Kaiping Peng, "Culture and Systems of Thought:
Holistic vs. Analytic Cognition," *Psychological Review* 108 (2001), pp. 291–310; Ulrich
Kühnen, Bettina Hannover, Ute Roeder, Ashiq Ali Shah, Benjamin Schubert, Arnold
Upmeyer, and Saliza Zakaria, "Cross-Cultural Variations in Identifying Embedded
Figures: Comparisons from the United States, Germany, Russia, and Malaysia," *Jour-
nal of Cross-Cultural Psychology* 32, no. 3 (2001), pp. 366–72; Marina Abalakina-Paap,
"Memory for Cultural Information about Russia and the United States: Effects of Na-
tionality and Mood," *Journal of Cross-Cultural Psychology* 32, no. 1 (2001), pp. 32–42;
Bernice Rosenthal, *The Occult in Russian and Soviet Culture* (Ithaca, NY: Cornell UP,
1997), p. 155.

97. Manfred F. R. Kets de Vries, "The Anarchist Within: Clinical Reflections
on Russian Character and Leadership Style," *Human Relations* 54, no. 5 (2001), pp.
585–627; Bathurst, *Intelligence and the Mirror*, pp. 18–31, 118–21; Raymond Cohen,
Negotiating across Cultures: International Communication in an Independent World
(Washington, DC: United States Institute of Peace Press, 1997), pp. 30–33; Glenn
and Glenn, pp. 80–81; Bonham, Sergeev, and Parshin, "Limited Test-Ban Agreement," p. 224.

98. Grossmann, "Iron Curtain," pp. 8, 14; Cote, *Russian Psychology in Transition*,
p. 84.

99. Bonham, Sergeev, and Parshin, "Limited Test-Ban Agreement."

100. For examples of the "holistic" and "organic" approach, see: Viktor Sergeev,
Tipy konceptualnykh interpretatzyi politicheskih peremen (Tartu: Izdatelstvo Tartusk-
ogo Universiteta, 1987).

101. According to Russian intellectual tradition, every notion was artificial and
disintegrated without the deeper insight of an all-embracing view according to which
every element was in constant interplay with others in frames of an indispensable and
synthetic metasystem. Isaiah Berlin, *The Soviet Mind: Russian Culture under Communism* (Washington, DC: Brookings Institution Press, 2004), pp. 132–34.

102. I. A. Il'in, *Put' k ochevidnsoti* (Moscow: AST, 1993), pp. 340–44; D. I. Mendeleev, *Zavetnye mysli* (Moscow: Mysl', 1995); Shapovalov, *Istoki i smysl rossiskoi civilizavii*, pp. 259–67; Lev Gumilevskii, *Russkie inzhenery* (Moscow: Molodaia gvardiia, 1953), pp. 110–15, 120–29; S. V. Solov'ev, *Nacionalnyi vopros v Rossii* (Moscow: AST, 1988), pp. 5–6; V. I. Vernadskii, *Nauchnaia mysl' kak planetnoe iavlenie* (Moscow: Nauka, 1991), pp. 9–10; N. I. Kareev, *Dukh russkoi nauki* (Moscow: Russkaia ideia, 1992), pp. 171–72; A. I. Gercen, "Diletantism v nauke," in *Izbrannye filisofskie proizvedeniia* (Moscow: Gospolitizdat, 1946), vol. 1, pp. 63–65; Vadim Kozhinov, *O russkom nacional'nom soznanii* (Moscow: Algoritm, 2002), pp. 371–73.

103. I. A. Il'in, *O Russkoi Idee* (Munich: Im Werden Verlag, 2006), pp. 1–8; S. Anisiomov, *Puteshestviia Kropotkina* (Moscow: Akademiia Nauk, 1943), pp. 112–13; K. A. Timiriazev, *Sobranie sochinenii* (Moscow: Akademiia Nauk, 1938), vol. 5, pp. 40–45. In his work on Russian science, Gumilevskii shows how this inclination toward a holistic, synthetic, meta-theory approach manifested itself in Russian physics, chemistry, mathematics, linguistics, and history. Gumilevskii, pp. 110–15, 120–29; Shapovalov, *Istoki i smysl rossiskoi civilizavii*, pp. 259–67; L. N. Gumilev, *Drevniaia Rus' I Velikaia step'* (Moscow: AST, 1989), pp. 19–20; Solov'ev, *Nacionalnyi vopros*, pp. 5–6; Vernadskii, *Nauchnaia mysl'*, pp. 9–10; Kareev, *Dukh nauki*, pp. 171–72.

104. Berlin, *Soviet Mind*, p. 134.

105. During the Soviet period prominent scientists promoted the merger of scientific disciplines for the sake of developing a metacosmic theory of science. G. P. Aksenov, *Vladimir Vernadskii: zhinzeopisanie i izbrannie trudy* (Moscow: Sovremennik, 1993). Michael Hagemeister, "Russian Cosmism in the 1920s and Today," in Bernice Rosenthal, *The Occult in Russian and Soviet Culture* (Ithaca, NY: Cornell UP, 1997), pp. 185–203.

106. Norbert Wiener, *Cybernetics* (Cambridge, MA: MIT Press, 1948).

107. Ludwig von Bertalanffy, *General System Theory: Foundations, Development, Applications* (New York: George Braziller, 1968).

108. A. A. Bogdanov, *Tektologiia—Vseobschaia organizatcionnaia nauka* (Moscow: Ekonomika, 1989).

109. Loren Graham, *Science in Russia and the Soviet Union* (New York: Cambridge UP, 1993), and *Science and the Soviet Social Order* (Cambridge, MA: Harvard UP, 1990), and *Science, Philosophy, and Human Behavior in the Soviet Union* (New York: Columbia UP, 1987), chs. 4 and 8.

110. Berlin, *Soviet Mind*, pp. 134–35, 142–43, 148–49.

111. Christopher Donnelly, *Red Banner: The Soviet Military System in Peace and War* (London: Jane's Information Group, 1988), p. 31.

112. Nathan Leites, *The Operational Code of Politburo* (New York: McGraw Hill, 1951), pp. 3–4; Bathurst, *Intelligence and the Mirror*, pp. 18–20, 126–27; also see Naveh, *In Pursuit of Military Excellence*, pp. 164–91.

113. The concept of operational maneuver, codified later in the "deep battle" theory, consisted of three main parts: the "fragmentation strike," which isolated an enemy's military subsystem from the strategic supersystem; "simultaneity," which combined and coordinated the actions of the frontal echelon with the deep operations in the enemy rear, in order to paralyze an enemy's system; and "momentum"—exploitation of the synergetic effect produced by fragmentation and simultaneity, in order to deny the opposing system time to respond. Naveh, *In Pursuit of Military Excellence*, ch. 1, pp. 214–36, 322–31.

114. Hans Melberg, "Four Distinctive Aspects of Soviet and Russian Military Thinking," http://www.geocities.com/hmelberg/papers/930928.htm; Yakov Falkov, *The Experience of Latvian HQ of the Soviet Partisan Movement, 1943–1944* (master's thesis, Tel Aviv University, 2005); Stuart A. Whitehead, "Balancing Tyche: Nonlinearity and Joint Operations," in Williamson Murray, *National Security Challenges for the 21st Century* (Carlisle, PA: Strategic Studies Institute, US Army War College, 2003), pp. 38–40.

115. Bonham, Sergeev, and Parshin, "Limited Test-Ban Agreement"; V. Sergeev, V. Akimov, V. Lukov, and P. Parshin, "Interdependence in a Crisis Situation: A Cognitive Approach to Modeling the Caribbean Crisis," *Journal of Conflict Resolution* 34, no. 2 (June 1990), pp. 179–207.

116. V. V. Kruglov, "Novyi podhod k analizy sovremennogo protivoborstva," *VM*, no. 12 (December 2006).

117. Shapovalov, *Istoki i smysl rossiskoi civilizavii*, pp. 214–17; F. Stepun, *Misli o Rossii: russkaia filosofiia sobstvennosti* (Moscow: ROSSPEN, 2000), pp. 336–37; Il'in, *Suschnost' i svoeobrazie*, p. 172; Zaicev, *Nacionalnoe*, pp. 32–34, and "Universlanoe," pp. 87–88; Mikhail Epstein, *Na granitzakh kultur: Rossiiskoe, Amerikanskoe, Sovetskoe* (New York: Slovo, 1995), p. 174.

118. Bathurst, *Intelligence and the Mirror*, pp. 16–18, 83.

119. David R. Jones, "Soviet Strategic Culture," in Carl Jacobsen, *Strategic Power: USA/USSR* (New York: St. Martin's Press, 1990), pp. 35–50; Il'in, *Put' k ochevidnsoti*; Zaicev, *Nacionalnoe*; Shapovalov, *Istoki i smysl rossiskoi civilizavii*, pp. 56–88. "If America is about success, Russia is about survival." Citation from Richard Lourie, *Predicting Russia's Future* (Knoxville, TN: Whittle Direct Books, 1991), p. 2.

120. Vladimir Mezhenkov, *Russkie: istoki, psihologia, sud'ba* (Moscow: Russkaia Kniga, 2003), pp. 155–57; Evgenii Evtushenko, "Preterpelost," *Literaturania gazeta*, 13 May 1988; Epstein, *Na granitzakh kultur*, p. 226; Grossmann, "Iron Curtain," p. 10.

121. Ksenia Kasianova, *O russkom nacional'nom kharaktere* (Moscow: Akademicheskii Proekt, 2003), pp. 476–77, 486–88; A. Andreev and A. Selivanov, *Russkaia tradicia* (Moscow: Algoritm, 2004), pp. 98–100; P. A. Sorokin, *Osnovnye cherty russkoi nacii v XX stoletii* (Moscow: Algoritm, 1990), pp. 471–72; K. D. Kavelin, *Nash umstvennyi stroi* (Moscow: RossPress, 1989), pp. 201–3; Kets de Vries, "The Anarchist Within";

S. Ia. Matveeva and V. E. Shlapentokh, *Strakhi v Rossii v proshlom i nastoiaschem* (Novosibirsk: Sibirskii Khronograph, 2000), p. 101.

122. Kasianova, *O russkom nacional'nom kharaktere*, pp. 476–77, 486–88; Ronald Inglehart and Hans-Dieter Klingemann, *Culture and Subjective Well-Being* (Cambridge, MA: MIT Press, 2000), pp. 165–83, cited in Grossmann, "Iron Curtain"; Richmond, *From Nyet to Da*, p. 129.

123. Bathurst, *Intelligence and the Mirror*, pp. 36–38, 110.

124. Ermarth, *Russia's Strategic Culture*, p. 5.

125. Bathurst, *Intelligence and the Mirror*, pp. 36–38, 110.

126. William C. Fuller, *Strategy and Power in Russia, 1600–1914* (New York: Free Press, 1992), chs. 1–6; Rice, "Soviet Strategy," pp. 659–63.

127. Ermarth, *Russia's Strategic Culture*, p. 5.

128. A. V. Suvorov, *Nauka pobezhdat* (Moscow: Molodaia gvardiia, 1984); G. P. Mescheriakov, *Russkaia voennaia mysl' v 19 veke* (Moscow: Voenizdat, 1973), pp. 198–201, 215–18; M. I. Dragomirov, *Izbrannye trudy* (Moscow: Voenizdat, 1956), p. 603.

129. Fuller, chs. 4–8; Walter Pintner, "Russian Military Thought: The Western Model and the Shadow of Suvorov," in Peter Paret, *Makers of Modern Strategy from Machiavelli to the Nuclear Age* (Princeton, NJ: Princeton UP, 1986), pp. 354–68.

130. Pintner, "Russian Military Thought," pp. 373–75; Mescheriakov, pp. 198–201, 215–18; Bathurst, *Intelligence and the Mirror*, pp. 36–38, 110.

131. Fuller, chs. 4–8; Pintner, "Russian Military Thought," pp. 372–73; Ivan Bliokh, *Budushchaia voina v tekhnicheskom, ekonomicheskom i politicheskom otnosheniiakh* (St. Petersburg, 1898); G. P. Mescheriakov, *Russkaia voennaia mysl' v 19 veke* (Moscow: Voenizdat, 1973), pp. 200–202, 246–47; Dragomirov, *Izbrannye trudy*, p. 603; L. G. Beskrovnyi, *Otcherki voennoi istoriografii Rossii* (Moscow: Nauka, 1962); A. Ageev, "Voenno teoriticheskie vzgliady N. P. Mihnevicha," *VIZh*, no. 1 (1975), pp. 90–94.

132. Rice, "Soviet Strategy," pp. 656–60; M. V. Frunze, *Sobranie sochinenii* (Moscow: Voen Izdat, 1929), pp. 253–55; Bathurst, *Intelligence and the Mirror*, p. 70.

133. Ermarth, *Russia's Strategic Culture*, p. 5; Mescheriakov, *Russkaia voennaia mysl' v 19 veke*; N. V. Medem, *Taktika* (St. Petersburg, 1837); M. I. Dragomirov, *Uchebnik taktiki* (St. Petersburg, 1879), cited in Pintner, "Russian Military Thought," pp. 357–58.

134. Walter Pintner, "The Burden of Defense in Imperial Russia," *Russian Review* 43, no. 3 (1984); A. A. Kilichenkov, "T-34 protiv Panzervaffe: iz istorii mental'nogo protivoborstva na sovetsko-germanskom frone (1941–1945)," *Novyi istoricheskii vestnik*, 1 September 2005.

135. V. V. Kirilov, "Chelovecheskii faktor kak osnovopologaiushii element voennoi moschi gosudarstva," *VM*, no. 4 (2006), pp. 17–24.

136. Catherine Merridale, *Ivan's War: Life and Death in the Red Army, 1939–1945* (New York: Metropolitan Books, 2006); Bathurst, *Intelligence and the Mirror*, ch. 3;

Nathan Leites, *Soviet Style in War* (New York: Crane, Russak, 1982), pp. 217–21, and *Operational Code*, p. 50; Dwight Eisenhower, *Crusade in Europe* (Garden City, NY: Doubleday, 1984), pp. 4–5; Miriam D. Becker, "Strategic Culture and Ballistic Missile Defense: Russia and the United States," *Airpower Journal* (special ed. 1994), pp. 5–6. For an alternative view, see: Amnon Sella, *The Value of Human Life in Soviet Warfare* (London: Routledge, 1992).

137. Kilichenkov; Ermarth, *Russia's Strategic Culture*, p. 5.

138. Jones, "Soviet Strategic Culture"; Kasianova, *O russkom nacional'nom kharaktere*; Bathurst, *Intelligence and the Mirror*; Richmond, *From Nyet to Da*, p. 129.

139. Donnelly, *Red Banner*, pp. 123–25; V. A. Rodin and O. Iu. Efremov, "O razvitii sistemy chelovek-voennaia tekhnika," *VM*, no. 3 (2007), pp. 61–65; "Voennaia promyshlennost,'" *Sovetskaia voennaia entciklopediia* (Moscow: Voenizdat, 1979), vol. 8; A. P. Gorkin, V. A. Zolotarev, and V. M. Karev, eds., *Voennyi entsiklopedicheskii Slovar'* (Moscow: Institut Voennoi Istorii, MO RF, 2001), pp. 390–94; N. S. Simonov, *Voenno-promyshlennyi kompleks v SSSR: tempy ekonomicheskogo rosta, struktura, organizatciia upravelniia* (Moscow: Simonov Press, 1996).

140. Donnelly, *Red Banner*, pp. 61–65, 125–34.

141. Vitalii Moroz, "Bitva na kul'manakh," *KZ*, 14 March 2007; Andrei Kokoshyn, "Otvet Rossii budet assimetrichtnyi," *KZ*, 7 April 2007; Oleg Grinevskii, *Stsenarii dlia tret'ei mirovoi* (Moscow: Olma Press, 2002).

142. Ogarkov, *Istoriia uchit bditel'nosti*, pp. 32–35.

143. Donnelly, *Red Banner*, p. 207.

144. Iu. Lotman and B. Uspenskii, "Rol' dualnykh modelei v dinamike Russkoi kultury," in Boris Uspenskii, *Semiotika istorii. Semiotika kultury* (Moscow: Izdatelstvo Gnozis, 1994).

145. Berlin, *Soviet Mind*, pp. 130–32.

146. During the 1920s in the Soviet Union, innovations took place across the board, varying from social reorganization in agriculture, to avant-gardism in art and poetry, to futuristic innovations in military science. Berlin, *Soviet Mind*, pp. 135–37, 139–40; Nikolai Bukharin, *Revoliutcia i kultura* (Moscow: Fond im. Bukharina, 1993).

147. A. D. Siniavskii, *Osnovy sovetskoi tsivilizatsii* (Moscow: Agraf, 2002), pp. 11–12, 55–57, 63–65, 75; Bathurst, *Intelligence and the Mirror*, pp. 38, 48–49; Shapovalov, *Istoki i smysl rossiskoi civilizavii*, pp. 41–43.

148. Siniavskii, *Osnovy sovetskoi tsivilizatsii*, pp. 11–12, 55–57, 63–65, 75; Bathurst, *Intelligence and the Mirror*, pp. 38, 48–49, 15, 87.

149. M. V. Frunze, *Edinaia veonnaia doktrina i Krasnaia Armiia* (Moscow: Voenizdat, 1921) and *Izbrannie proizvedeniia* (Moscow: Voenizdat, 1977), pp. 27–49; S. I. Gusev, *Grazhdanskaia voina i Krasnaia Armiia* (Moscow: Voenizdat, 1958), pp. 210–

30; Rice, "Soviet Strategy," pp. 654–55; John Erickson, *The Soviet High Command: A Military Political History* (New York: St. Martin's Press, 1962), pts. 1 and 2; Jacob W. Kipp, *Mass, Mobility, and the Red Army's Road to Operational Art, 1918–1936* (Fort Leavenworth, KS: Command and General Staff College, 1988), p. 16.

150. Rice, "Soviet Strategy," pp. 659–60; Tukhachevsky suggested that military science was an area where political interference or sensitivity to the current political line could be damaging. M. N. Tukhachevsky, "Voina kak problema vooruzhennoi bor'by," in *Boevoi put' Sovetskikh vooruzhennykh sil* (Moscow: Voenizdat, 1960).

151. M. V. Frunze, *Izbrannie proizvedeniia* (Moscow: Voenizdat, 1977), pp. 196–99.

152. Donnelly, *Red Banner*, pp. 100–103; Ghulam Wardak and Graham Turbiville Jr., eds., *Voroshilov Lectures* (Washington, DC: National Defense UP, 1989), vol. 1, pp. 3–4.

153. F. F. Gaivoronovskii and M. I. Galkin, *Kultura voennogo myshlenia* (Moscow: Voenizdat, 1991).

154. V. K. Konoplev, *Nauchnoe predvidenie v voennom dele* (Moscow: Voenizdat, 1974), pp. 6, 13, 32–33, 65–70, 127; S. I. Krupnov, *Dialektika i voennaia nauka* (Moscow: Voenizdat, 1963), pp. 100–126; I. A. Grudinin, *Dialektika i sovremennoe voennoe delo* (Moscow: Voenizdat, 1971); Ogarkov, *Vsegda v gotovnosti*, pp. 41–45; V. M. Bondarenko, "Nauka kak factor ukrepleniia oboronsoposobnosti strany," in A. S. Milovidov, *Voenno-teoreticheskoe nasledie V. I. Lenina i problemi sovremennoi voini* (Moscow: Voenizdat, 1987).

155. Gareev, *Elsi zavtra*; Ogarkov, *Istoriia uchit bditel'nosti*.

156. A. A. Kokoshin, *Inovatcionnye vooruzhennye sily i revoliutciia v voennom dele* (Moscow: URSS, 2009); Bondarenko, *Sovremennaia nauka*, pp. 109–11; I. Anureev, "Nauchno tekhnicheskii progress i voennaia nauka," *VM*, no. 2 (1970), pp. 27–39; M. Cherednichenko, "Nauchno tekhnicheskii progress i razvitie vooruzhenia i voennoi tekhniki," *VM*, no. 4 (1972), pp. 29–41. Otuytsskiy, "K Voprosu o Sushchnosti voenno-tekhinicheskikh revoliutsii," pp. 52–59; Derevianko, *Revoliutsiia v voennom dele*.

157. "Prognozirovanie" and "predvidinie" entries in *Voennyi Entsiklopedicheskii Slovar'* (Moscow: Voenizdat, 1983), p. 585; Shavrov and Galkin, *Metodologiia voenno-nauchnogo poznaniia*.

158. A. O. Baranov, *Suschnost' i soderzhanie sovremennoi voenno tekhnicheskoi revoliutzii* (Moscow: Ministerstvo Oborony, 1977); N. A. Lomov, *Nauchno tekhnicheskii progress i revoliutzia v voennom dele* (Moscow: Ministerstvo Oborony, 1973); Bondarenko, *Sovremennaia nauka i razvitie voennogo dela*; Gaivoronovskii, *Evoliutzia voennogo iskusstva*.

159. S. Kozlov, "K voprosu o razvitii sovetskoi voennoi nauki posle vtori mirovoi voiny," *VM*, no.2 (February 1964), pp. 64–73. According to dialectical laws, future developments were inevitable, and none of fundamental trends was considered acciden-

tal. This determinism enhanced the belief in the ability to predict events scientifically. Leites, *Operational Code*, pp. 1–4, 15–16.

160. The Soviets took a sophisticated, if at times contradictory, philosophical approach to forecasting the future. As opposed to a military theoretican, a Soviet tactical commander prepared himself for future "variants" by learning a certain number of battle drills that simulated all possible scenarios. Uncontrolled improvisation was forbidden. The basic direction for heightening readiness for operational surprise was autoimmunization of the system of action that has to be performed. This mechanical learning did not contradict the call for creative analysis of the situation on the level of military thought. Eisenstadt and Pollack, pp. 69–70; Leites, *Soviet Style in War*, chs. 3 and 6, pp. 330–48; F. F. Gaivoronovskii and M. I. Galkin, *Kultura voennogo myshlenia* (Moscow: Voennizdat, 1991); I. E. Shavrov and M. I. Galkin, *Metodologiia voenno-nauchnogo poznaniia* (Moscow: Voenizdat, 1977).

161. I. E. Shavrov and M. I. Galkin, eds., *Metodologiia voenno-nauchnogo poznaniia*, pts. 1, 3, 6; Gaivoronovskii and Galkin, *Kultura voennogo myshlenia*, chs. 2 and 4. On the army and front levels, Soviet military theory called for creativity and improvisation (*operativnoe tvrochestvo*). Naveh, *In Pursuit of Military Excellence*, pp. 235–36.

162. Strong central leadership—the czar, a nobleman, a landowner, Lenin, Stalin, or a Communist Party commissar—has been perceived by Russians as necessary for holding the country together. Manfred F. R. Kets de Vries, I. Gurkov, and Y. Kuz'minov, "Organizational Learning in Russian Privatized Enterprises," *International Studies in Management and Organization* 25, no. 4 (1995), pp. 91–117.

163. Kets de Vries, Gurkov, and Kuz'minov, "Organizational Learning in Russian."

164. Bathurst, *Intelligence and the Mirror*, pp. 110–11; Evangelista, *Innovation and the Arms Race.*

165. Boris Shaposhnikov, *Mozg Armii* (Moscow: Voennaia Literatura, 1978); M. V. Zakharov, *General'nyi shtab v predvoennye gody* (Moscow: Voenizdat, 1989).

166. A. A. Kokoshin, *Strategicheskoe upravelnie* (Moscow: ROSSPEN, 2003), pp. 351–66; S. L. Pechurov, *Anglo-saksonskaia model' upravleniia v voennoi sfere* (Moscow: KomKniga, 2005), pp. 162–70; I. E. Shavrov, "Voennaia Akademiia General'nogo Shtaba," in Ogarkov, *Entziklopedia*; V. G. Kulikov, *Akademiia General'nogo Shtaba* (Moscow: Voenizdat, 1976).

167. Aleksandr Savinkin, "A. Svechin: zadachi general'nogo shtaba i intelektualism v voennom vospitanii," *Nezavisimoe voennoe obozrenie*, 17 April 1998; see Aleksandr Svechin, *Strategia* (Moscow: Voennyi Vestnik, 1927) and *Evoliutcia voennogo iskusstva* (Moscow: Kuchkovo Pole, 2002).

168. Donnelly, *Red Banner*, pp. 144–45.

169. Wardak and Turbiville Jr., *Voroshilov Lectures*, pp. 23–40.

170. Makhmut Gareev, "Vydaiuschiisia voennyi reformator," *KZ*, 30 October 2002; Aleksandr Kochiukov, "Strateg i politik," *KZ*, 31 October 2002; Varennikov,

Nepovtorimoe. General Sergei Shtemenko, the legendary chief of the main operational directorate, who planned the strategic operations of WWII together with Stalin, served most of his life as an officer for operational planning, without any combat experience whatsoever. Aleksandr Skvortzov, "Strateg kazatchego roda," *KZ*, 20 February 2007.

171. Steven Meyer, "Decision Making in Defense: The Soviet Case," in Jacobsen, *Strategic Power*, pp. 247–60.

172. Fedor Kozanchuk, "Novye knigi: kvintesencia voennykh znanii," *Rossiiskoe voennoe obozrenie*, 31 March 2005.

173. "Revoliutciia v voennom dele," *Sovetskaia voennaia entciklopediia* (Moscow: Voenizdat, 1979), vol. 7.

174. Wardak and Turbiville Jr., *Voroshilov Lectures*, pp. 23–40.

175. Stalin argued in 1924 that the party must combine the broad outlook of the Russian revolutionary with American practicality. Leites, *Operational Code*, p. 30.

176. Richmond, *From Nyet to Da*, pp. 128–30.

177. Donnelly, *Red Banner*, p. 29; Andreev and Selivanonov, *Russkaia tradicia*, pp. 45–48; Bathurst, *Intelligence and the Mirror*; L. P. Krasavin, "Vostok, Zapad i russkaia idea," in Arsenii Gulyga, *Russkaia ideia: sbornik proizvedenii russkikh myslitelei* (Moscow: Airis Press, 2004), pp. 318–51; Cote, *Russian Psychology in Transition*, p. 196; Grossmann, "Iron Curtain," p. 15.

178. Bondareva, pp. 50–51; Mezhenkov, *Russkie*, pp. 140–43; Howard F. Stein, "Russian Nationalism and the Divided Soul of the Westernizers and Slavophiles," *Ethos* 4, no. 4 (Winter 1976), pp. 403–38; Shapovalov, *Istoki i smysl rossiskoi civilizavii*; Gumilev, *Drevniaia Rus'*. While Americans tend to extrapolate from the present situation to a kind of permanent future, the Russians project the future onto the present; Bathurst, *Intelligence and the Mirror*, pp. 112–13.

179. Bathurst, *Intelligence and the Mirror*.

180. Neidhart, *Russia's Carnival*, pp. 63–71.

181. Richmond, *From Nyet to Da*, p. 126; Stephen Blank, *Potemkin's Treadmill: Russian Military Modernization* (National Bureau of Asian Research, 2005); Neidhart, *Russia's Carnival*, pp. 63–66; Bathurst, *Intelligence and the Mirror*. Soviet high culture and institutions were embodiments of "make-believe." They did not spring from reality, but rather created it. Bondareva, pp. 40–45. This avoidance of reality reached incredible heights under the State Planning Committee (GOSPLAN). Kets de Vries, Gurkov, and Kuz'minov, "Organizational Learning in Russian."

182. Scott W. Palmer, *Dictatorship of the Air: Aviation Culture and the Fate of Modern Russia* (Cambridge: Cambridge UP, 2006).

183. Fritz W. Ermarth, *Russia's Strategic Culture: Past, Present and . . . in Transition?* (SAIC report for the Defense Threat Reduction Agency, 31 October 2006), p. 6.

184. Donnelly, *Red Banner*, p. 38; Richmond, *From Nyet to Da*, pp. 26–33; Ermarth, *Russia's Strategic Culture*, pp. 6–7; Grossmann, "Iron Curtain," p. 9.

185. Dale Pesmen, *Russia and Soul: An Exploration* (Ithaca, NY: Cornell UP, 2000); Anna Wierzbicka, "Soul and Mind: Linguistic Evidence for Ethno-psychology and Cultural History," *American Anthropologist* 91 (1989), pp. 41–58; Il'in, *Sushcnost' i svoeobrazie*, pp. 178–89; N. A. Berdyaev, "Russkaia idea," in M. A. Maslin and A. L. Andreev, *O Rossii i russkoi filosofskoi kulture* (Moscow: Nauka, 1990), pp. 44–45, 90–92; Grossmann, "Iron Curtain," p. 7.

186. Andreev and Selivanov, *Russkaia tradicia*, pp. 70–74, and "Zapadnyi individualism i russkaia tradicia," *Filosofiia i obshestvo*, no. 4 (2001); S. Kara-Murza, *Sovetskaia tzivilizaciia* (Moscow: Olma Press, 2002); F. Stepun, *Misli o Rossii*, pp. 336–37; Il'in, *Sushcnost' i svoeobrazie*, pp. 178–89; Anna Wierzbicka, "Russian Cultural Scripts: The Theory of Cultural Scripts and Its Applications," *Ethos* 30, no. 4 (December 2002), pp. 401–32; Mikhail Epstein, *Na granitzakh kultur: Rossiiskoe, Amerikanskoe, Sovetskoe* (New York: Slovo, 1995), p. 87; Grossmann, "Iron Curtain," p. 11; A. Pavlenko, "Emotions and the Body in Russian and English," *Pragmatics and Cognition* 10, no. 1 (2002), pp. 207–41.

187. This sense of pessimistic impotence—an apathetic attitude toward environmental forces—is a prevalent theme throughout Russia's history. Kets de Vries, "The Anarchist Within" and "Organizational Learning."

188. Bathurst, *Intelligence and the Mirror*, pp. 31–33.

189. Andrei Sinevskii, *Osnovy sovetskoi tzivilizacii* (Moscow: Agraff, 2002), pp. 17–18, 49–52; Andrei Platonov, *Kotlovan* (Moscow: Shkola Press, 1995); Andreev and Selivanov, *Russkaia tradicia*, pp. 103–5; M. Mikheev, "Chuvstvo uma i myslimost' chiuvstva u Platonova," *Voprosy filosofii*, no. 7 (2001); A. Razin, "Russkaia ideia i sovrremennaia Rossia: krizis i perspektivi razvitija," *Filosofskie nauki*, no. 4 (2001); Mezhenkov, *Russkie*, pp. 144–45.

190. Richmond, *From Nyet to Da*, p. 130.

191. Kapitsa even discussed Gumilevsky's ideas in his correspondence with Stalin, trying to force the latter to invest more in the practical and applied aspects of science. Kozhinov, *O russkom nacional'nom soznanii*, pp. 371–73.

192. Epstein, *Na granitzakh kultur*, p. 110.

193. Graham, *Science, Philosophy, and Human Behavior*, ch. 8.

194. Bathurst, *Intelligence and the Mirror*, pp. 33–35, 84, 112–13; Leites, *Soviet Style in War*, pp. 1–4, 6–8, 15–17, and *Soviet Style in Management* (New York: Crane Russak, 1985); Donnelly, *Red Banner*, p. 29; Kokoshin, *Inovatcionnye vooruzhennye*, p. 15; G. G. Kostev, *Voenno-morskoi flot strany* (St. Petersburg: Nauka, 1997), pp. 463–64.

195. Bathurst, *Intelligence and the Mirror*, pp. 33–35, 82–84. These tendencies penetrated Soviet society so deeply and were so blatant that the party leadership openly acknowledged them and attempted to counterbalance them by enhancing activeness, increasing material stimulation, appealing to people's sense of duty, and elevating collective surge. Leites, *Soviet Style in Management*, ch. 1.

196. Melberg, "Four Distinctive Aspects."

197. Poverty is often considered to be the father of ingenuity, and strategic necessity the mother of military invention and creativity. Earl Tilford, *The Revolutions in Military Affairs: Problems and Cautions* (Strategic Studies Institute, US Army War College, 1995), pp. 11–12.

198. John Erickson, *Soviet Ground Forces: An Operational Assessment* (Boulder, CO: Westview, 1986). To a certain degree it contradicted the Soviet military tradition and the structure of the Soviet system as a whole. The operational implementation of MTR in the form of ROK/RUK and OMG assumed the allocation of unprecedented authority to the level of tactical commanders. However, the battle drills up to battalion level aimed to ensure that tactical commanders could not, through initiatives of their own, upset plans and options of senior commanders. Far from being mechanical robots, the Soviet tactical units' commanders, nonetheless, had a certain number of battle drills available for any given situation. They were expected to tailor a given battle drill to the mission, enemy, terrain, and other operational considerations. The tactical units naturally enjoyed a certain level of combat independence and were expected to execute flank maneuvers, to envelop an enemy's formations, to break through into the deep enemy rear, and to seize key objectives. However, Soviet military thought argued that wars were won at the operational, not the tactical, level. Consequently, resources and authority were concentrated at the operational levels, purposely limiting the freedom of tactical commanders. Michael Eisenstadt and Kenneth Pollack, "Armies of Snow and Armies of Sand: The Impact of Soviet Military Doctrine on Arab Militaries," in Emily Goldman and Leslie Eliason, *Diffusion of Military Technology and Ideas* (Stanford, CA: Stanford UP, 2003), pp. 69–70.

199. Christopher Donnelly, "The Soviet Operational Maneuver Group: A New Challenge for NATO," *International Defense Review* 15, no. 9 (1982).

200. In contrast to the 1930s, theoretical writings on tactics were absent from the military literature following the Great Patriotic War. V. K. Trinadofilov, *Kharakter operatccii sovremennykh armii*; A. I. Verhovskii, *Obschaia taktika*; *Ogon,' manevr, maskirovka*. V. Reznichenko, *Taktika* (Moscow: Voenizdat, 1966, 1984, 1987).

201. I. N.Vorob'ev, "Pochemu taktika okazalas' v zastoe," *VM*, no. 1 (1989), "O tvorchestve i novatarstve v taktike," *VM*, no. 3 (1988), and "Takticheskaia oborona," *VM*, no. 1 (1989); A. M. Adgamov, "Protiv shablona i konservatizma v taktike i takticheskoi podgotovke," *VM*, no. 10 (1988); A. P. Kurkov, "O svobode tvorchestva i nauchnom potenciale professorsko-prepodavatelskogo sostava akademii," *VM*, no. 11 (1986); A. F. Shramchenko, "Vyrabotka tvorcheskogo myshleniia u slushatelei v khode operativno-takticheskoi podgotovki," *VM*, no. 3 (1990); M. A. Gareev, "Razvitie form i metodov."

Chapter 3

1. Jeffrey McKitrick, "The Revolution in Military Affairs," in *Battlefield of the Future: 21st Century Warfare Issues* (Maxwell AFB, AL: Air UP, 1995).

2. Andrew W. Marshall, *Some Thoughts on Military Revolutions—Second Version* (Washington, DC: Office of the Secretary of Defense, 1993), p. 1.

3. William Owens, *Lifting the Fog of War* (New York: Farrar Straus Giroux), p. 82.

4. Robert Tomes, *Military Innovation and the Origins of the American Revolution in Military Affairs* (PhD diss., Department of Government and Politics, University of Maryland, 2004), pp. 200–208.

5. William Perry, testimony to the US Senate Armed Services Committee, Hearing on DoD Appropriations for FY1977, in Vickers and Martinage, *Revolution in War*, pp. 8–9; also see: Owens, *Lifting the Fog*, p. 81.

6. Richard H. Van Atta, Seymour J. Deitchman, and Sidney G. Reed, *DARPA Technical Accomplishments, Volume 3* (Alexandria, VA: Institute for Defense Analyses, 1991), pp. 2–14; Vickers and Martinage, *Revolution in War*, p. 9.

7. Owens, *Lifting the Fog*, pp. 82–83. According to Perry, the post–Cold War advances in US military effectiveness were an outgrowth of this strategy. William J. Perry, *Preventive Defense: A New Security Strategy for America* (Washington, DC: Brookings Institute Press, 1999), pp. 179–80.

8. Tomes, *Military Innovation*, pp. 154–221; Richard Van Atta and Michael Lippitz, *Transformation and Transition: DARPA's Role in Fostering the Emerging Revolution in Military Affairs* (Alexandria, VA: Institute for Defense Analyses, 2003), pp. 1–56; Vickers and Martinage, *Revolution in War*, pp. 8–10; Barry Watts, *Six Decades of Guided Munitions and Battle Networks: Progress and Prospects* (Washington, DC: CSBA, 2007).

9. Tomes, *Military Innovation*, pp. 154–215, 225.

10. Tomes, *Military Innovation*, pp. 242, 297 (chs. 4–5); Richard Van Atta, Jack Nunn, and Alethia Cook, "Assault Breaker," in Richard H. Van Atta et al., *DARPA Technical Accomplishments, Volume 2—Detailed Assessments* (Alexandria, VA: Institute for Defense Analyses, 2003), pp. iv–41.

11. Shimon Naveh, *In Pursuit of Military Excellence: The Evolution of Operational Theory* (London: Frank Cass, 1997).

12. John Erickson, "The Significance of Operational Art and the Development of Deep Battle," in John Gooch, *The Origins of Contemporary Doctrine* (Camberley, UK: Strategic and Combat Studies Institute, 1997), pp. 106–7.

13. Erickson, "Significance of Operational Art"; Richard Lock-Pullan, *US Intervention Policy and Army Innovation* (London: Taylor and Francis, 2006); David Glantz, *The Evolution of Soviet Operational Art, 1927–1991* (London: Frank Cass, 1995).

14. Lock-Pullan, *US Intervention*.

15. Tomes, *Military Innovation*, pp. 13, 16, 18, 255–56. This rediscovery of the campaign, the orchestration of theater military activities and planning conventional warfare at the operational level led to a renaissance of American military thought. The deputy commander of the US TRADOC saw the adoption of operational art as the most important change in army doctrine since World War II. L. D. Holder, "A New Day for Operational Art," in R. L. Allen, *Operational Level of War—Its Art* (Carlisle, PA: US Army War College, 1985).

16. Saul Bronfeld, "Fighting Outnumbered: The Impact of the Yom Kippur War on the U.S. Army," *Journal of Military History* 71, no. 2 (April 2007), pp. 465–98.

17. Lock-Pullan demonstrates that the Soviet conception of maneuver and firepower presented in the Soviet "deep-battle" works was the source of inspiration and emulation for many ALB concepts and was eventually adapted to US military thinking. Lock-Pullan, *US Intervention*, p. 93. US thought also borrowed from Soviet concepts of the 1970s, such as "operational maneuvering group." Tomes, *Military Innovation*, p. 301.

18. According to Erickson, the 1982 FM 100–5, by adopting the principle of equal importance of firepower and maneuver, and by distinguishing tactics from operations, along with its later move toward operational art, was clearly indebted to Soviet military thinking. Erickson, "Development of Soviet Military Doctrine," in John Gooch, ed., *The Origins of Contemporary Doctrine* (Camberley, UK: Strategic and Combat Studies Institute, 1997), pp. 106–7.

19. Tomes, *Military Innovation*, p. 265.

20. Remarks made by Prof. Azar Gat to author, 28 October 2008.

21. Thomas Mahnken, *Technology and the American Way of War* (New York: Columbia UP, 2008), p. 2.

22. William Perry, "Desert Storm and Deterrence," *Foreign Affairs*, no. 70 (Fall 1991), pp. 66–82.

23. Paul Dickson, *Sputnik: The Shock of the Century* (New York: Walker, 2001), p. 194; also see Van Atta and Lippitz, *Transformation and Transition*, vol. 1, pp. 1–2; Columba Peoples, "Sputnik and 'Skill Thinking' Revisited: Technological Determinism in American Responses to the Soviet Missile Threat," *Cold War History* 8, no. 1 (February 2008), pp. 55–75.

24. Tomes, *Military Innovation*, p. 323.

25. Owens, *Lifting the Fog*, pp. 82–83. In fact, operational cognition coevolved with perceptions of the RMA and expectations about information technology applied to battlefield problems. The ALB demonstrated that the level of American military thinking became far more sophisticated as it made the transition from sequential annihilation to understanding combat in terms of the broader operational level. Lock-Pullan, *US Intervention*, p. 685.

26. Perry, "Desert Storm and Deterrence."

27. Despite the degree of success, which RAND analysts called "spectacular," the US Air Force basically turned its back on precision-guided munitions until the Gulf War. It

was only in the aftermath of realizing how effective PGMs were in 1991 that the US AF committed itself institutionally to embracing that weaponry as the centerpiece of future strike operations. Correspondence with Barry Watts, October–November 2006; also see: Barry Watts, "American Air Power," in Williamson Murray, *The Emerging Strategic Environment: Challenges of the Twenty-first Century* (Westport, CT: Praeger, 1999), pp. 183–218; George and Meredith Friedman, *The Future of War: Power, Technology, and American World Dominance in the 21st Century* (New York: Crown, 1996), pp. 237–40.

28. See the March/April issues of *Survival*, 1979; Edward Luttwak, "American Style of Warfare and the Military Balance," considered the impact of the PGMs on the maneuver-attrition balance; James Digby, "New Technology and Super-Power Actions in Remote Contingencies," considered the impact of new technologies on power projection; John Mearsheimer, "Precision-Guided Munitions and Conventional Deterrence," considered the impact of the PGMs on the defense–offense balance. Also see Samuel Huntington, "Conventional Deterrence and Conventional Retaliation in Europe," *International Security* 8, no. 3 (1983/84).

29. MacGregor Knox and Williamson Murray, *The Dynamics of Military Revolution, 1300–2050* (Cambridge, UK: Cambridge UP, 2001), p. 3; Murphy Donovan, "Strategic Literacy," *Air Power Journal* (Winter 1988).

30. National Foreign Assessment Center, SR 79–10338X, *Soviet Military Theory: Structure and Significance*, October 1979; CIA FOIA Electronic Reading Room (hereafter ERR), p. 6.

31. Deputy for national intelligence officers, to assistant chief of staff/intelligence, Department of the Air Force, *Soviet Military Thought*, 17 May 1974; deputy to the DCI for collection tasking to director of Central Intelligence, *Possible Reductions of Air Force Translation of Soviet Documents*, 21 August 1978; Gen. James Brown, asst. chief of staff, intelligence, Department of the Air Force, to director, Central Intelligence, *USAF Efforts in the Field of Literature Intelligence*, 21 November 1977, US National Archives and Records Administration (hereafter, NA).

32. FBIS, *War and the Army: A Philosophical and Sociological Study*, ed. D. A. Volkogonov, A. S. Milovidov, and S. A. Tyushkevich, JPRS L/9649, 7 April 1981, pp. 1–7, 16–17, 21, 24, 136, 141, 148, 167–71; FBIS, *Methodology of Military Scientific Cognition*, JPRS 1/8213, 11 January 1979, pp. 12–29, NA.

33. US Joint Publications Research Service, *Translations on USSR Military Affairs: Basic Military Training*, FOUO 11/79/ JPRS L/8421, 25 April 1979. For the reference to the MTR, see especially pp. 33–34 and 222; FBIS, *Translations from* Voyennaia Mysl', *No. 12, 1971*, FPD 0003/73, 17 January 1974, pp. 87–88; FBIS, *Translations from* Voyennaia Mysl', *No. 10, 1971*, FPD 0008/74, 11 February 1974, p. 6; FBIS, *Translations from* Voyennaia Mysl', *No. 7, 1971*, FPD 0014/74, 7 March 1974, pp. 1–3, 6, NA.

34. ACS/AF/Intelligence to deputy for national intelligence officers, *Soviet Military Thought Translation Series*, 13 May 1974, NA.

35. FBIS/USSR Report/Military Affairs, *Military Science, Theory, Strategy: Forecasting in Military Affairs, Vol. 6, 1978*, FOUO 1/1981, 26 March 1981, pp. 1–6; FBIS/Translations on USSR Military Affairs, *Sociological Study of the Soviet Military Engineer*, FOUO 3/79, especially pp. 396 and 408, NA.

36. National Foreign Assessment Center, SR 81–18935X, *The Development of Soviet Military Power: Trends since 1965 and Prospects for the 1980s*, 13 April 1981, p. 67, ERR; Nikolai Pushkarev, *GRU: Vymysly i real'nost'—spetsluzhba voennoi razvedki* (Moscow: Eksmo, 2004), pp. 121–27; Ivan Potapov, "Ot Khrushchiova do Gorbacheva," *KZ*, 11 February 2006.

37. National Foreign Assessment Center, SR 81–18935X, pp. 67–69; and Central Intelligence Agency Directorate of Intelligence, SW-86 20026DX, *Soviet Artillery Precision-Guided Munitions: A Conventional Weapons Initiative*, September 1986; Special National Intelligence Estimate, *Soviet Acquisition of Military Significant Western Technology*, September 1985, ERR.

38. Directorate of Intelligence, SOV 84–10173, *Soviet Ground Forces Trends*, 1 October 1984, pp. 19–20; and National Intelligence Estimate, NIE 11–14–79, *Warsaw Pact Forces Opposite NATO*, 31 January 1979, p. 78, ERR.

39. Central Intelligence Agency Directorate of Intelligence, SW 86–10062, *Soviet Microelectronics: Impact of Western Technology Acquisitions*, December 1986; and National Intelligence Estimate, NIE 11–12–83, *Prospects for Soviet Military Technology and Research and Development*, 14 December 1983, ERR.

40. Director of Central Intelligence, *Trends and Development in Warsaw Pact Theater Forces and Doctrine through the 1990s*, NIE 11–14–89, February 1989, ERR.

41. The Assault Breaker, designated by the Soviets as ROK, and envisioned as a pivot of conventional theater operation, was the 1978 DARPA project that leveraged emerging technology to foster significant change in command and control capabilities, mobility, armor, night fighting, massed firepower, and precision stand-off fire. Van Atta, Nunn, and Cook, "Assault Breaker," p. IV-14.

42. Directorate of Central Intelligence, NIE 11/20–6-84, *Warsaw Pact Nonnuclear Threat to NATO Airbases in Central Europe*, 25 October 1984, pp. 41–42; and National Intelligence Estimate, NIE 11–14–85/D, *Trends and Developments in Warsaw Pact Theater Forces, 1985–2000*, September 1985, pp. 9–13, 29–33, ERR.

43. Director of Central Intelligence, NIE 11–14–89.

44. Major General Shlipchenko, cited in the CIA Directorate of Intelligence, *The USSR: Initial Military Reaction to Desert Storm*, 26 February 1990, p. 3, ERR.

45. National Intelligence Estimate, NIE 11–14–79, p. 79; National Foreign Assessment Center, SR 81–18935X, pp. 67–69.

46. Statement by Andrew Marshall at CSBA roundtable on future warfare, 12 March 2002, in Vickers and Martinage, *Revolution in War*, p. 11; Andrew F. Krepinevich, *The Military-Technical Revolution: A Preliminary Assessment* (Washington, DC: Center for Strategic and Budgetary Assessments, 2002), p. i.

47. Derek Leebaert, *The Fifty-Year Wound: The True Price of America's Cold War Victory* (Boston: Little, Brown, 2002), p. 507; Ronald E. Powaski, *The Cold War: The United States and the Soviet Union, 1917–1991* (New York: Oxford UP, 1998), p. 233.

48. Other analysts were Mary C. FitzGerald, Notra Truelock, and experts at Andrew Marshall's Office of Net Assessment.

49. This failure of interpretation seemed to be in keeping with the US intelligence community's slow recognition of other fundamental Soviet issues, such as assessing the collapse of the USSR. David Arbel and Ran Edelist, *Western Intelligence and the Collapse of the Soviet Union, 1980–1990* (London: Frank Cass, 2003).

50. Knox and Murray, *Dynamics of Military Revolution*, p. 3.

51. Interview with Makhmut Gareev, Moscow, July 2006; also see: Sergei Modestov, "Serii Kardinal Pentagona Andrew Marshall—ideolog novoi amerikanskoi revoliucii v voennom dele," *Nezavisimoe voennoe obozrenie*, no. 4, 14 December 1995.

52. Knox and Murray, *Dynamics of Military Revolution*, p. 4; James Der Derian, *Virtuous War* (Oxford: Westview Press, 2001), pp. 29–32. The Gulf War was something like Cambrai.

53. Jeffrey R. Cooper, "Another View of the Revolution in Military Affairs," in John Arquilla and David Ronfeldt, *In Athena's Camp* (RAND: National Defense Research Institute, 1997), p. 139, n. 39.

54. Tomes, *Military Innovation*, p. 336.

55. Andrew Bacevich, *The New American Militarism* (Oxford: Oxford UP, 2005), pp. 161–63; Stephen Rosen, "Net Assessment as an Analytical Concept," in Marshall, *Some Thoughts on Military Revolutions*, pp. 283–84.

56. Bacevich, *New American Militarism*; Albert Wohlstetter, "Threats and Promises of Peace: Europe and America in the New Era," *Orbis* 17 (Winter 1974), "Between an Unfree World and None: Increasing Our Choices," *Foreign Affairs* 63 (Summer 1985), and "The Political and Military Aims of Offensive and Defensive Innovation," in Fred Hoffman, Albert Wohlstetter, and David Yost, *Swords and Shields: NATO, the USSR, and New Choices for Long-Range Offense and Defense* (Lexington, MA: Lexington Books, 1987); Alex Abella, *Soldiers of Reason* (Orlando, FL: Harcourt Books, 2008), p. 277.

57. Krepinevich, *Military-Technical Revolution*, pp. i–iv.

58. Fred C. Ikle and Albert Wohlstetter, *Discriminate Deterrence: Report of the Commission on Integrated Long-Term Strategy* (Washington, DC: DoD, 1988), pp. 8, 29, 49, 65; Bacevich, *New American Militarism*, pp. 160–62.

59. Notra in Marshall, *Some Thoughts on Military Revolutions*, p. 143.

60. Andrew W. Marshall and Charles Wolf, *The Future Security Environment*, Report of the Future Security Environment Working Group, submitted to the Commission on Integrated Long-Term Strategy (Washington, DC: DoD, 1988), pp. 34–35, 40, 42, 64, 69–71.

61. Stephen J. Blank, *The Soviet Military Views of Operation Desert Storm: A Preliminary Assessment* (Carlisle, PA: Strategic Studies Institute, US Army War College, 1991), pp. 31–33; Naveh, *In Pursuit*, pp. 238 and 330; Norman C. Davis, "An Information-Based Revolution in Military Affairs," in John Arquilla and David Ronfeldt, *In Athena's Camp* (RAND: National Defense Research Institute, 1997), p. 85; Cooper, "Another View," p. 124; A. A. Kokoshin, *O Politicheskom smysle pobedy v sovremmennoi voine* (Moscow: URSS, 2004), pp. 36–37; Edward Felker, *Russian Military Doctrinal Reform in Light of Their Analysis of Desert Storm* (Maxwell AFB, AL: Air UP, 1995), p. 33; Bacevich, *New American Militarism*.

62. Statement by Andrew Marshall at a CSBA roundtable on future warfare, 12 March 2002, in Vickers and Martinage, *Revolution in War*, p. 12; Krepinevich, *Military-Technical Revolution*, pp. i–iv.

63. Marshall, *Some Thoughts on Military Revolutions*, pp. 2–4; Krepinevich, *Military-Technical Revolution*, pp. iii–iv, 5–7; Vickers and Martinage, *Revolution in War*, pp. 10–13; Michael Horowitz and Stephen Rosen, "Evolution or Revolution?" *Journal of Strategic Studies* 3 (June 2005), pp. 439–40.

64. According to the report's authors, changes in two fields (i.e., concepts of operations and organizational structures) would enable exploiting the technologies available now in the most revolutionary way. Krepinevich outlined a guiding principle for future defense reform: to integrate information networks (reconnaissance, surveillance, tracking, target acquisition, and BDA) with networks of weapon systems. These two fields would later encompass the four main components of the American RMA: precision strike, information dominance, space warfare, and dominant maneuver. Krepinevich, *Military-Technical Revolution*, p. 8; Marshall, *Some Thoughts on Military Revolutions*, pp. 2–4.

65. Marshall, *Some Thoughts on Military Revolutions*, p. 2.

66. Krepinevich, *Military-Technical Revolution*, p. iv; Cooper, "Another View," p. 135, n. 1. See for the Soviet "use" of the *RMA* term: P. M. Derevianko, *Revoliutsiia v voennom dele: vchem ee sushchnost'?* (Moscow: Ministerstvo Oborony SSSR, 1967); V. M. Bondarenko, *Sovremennaya nauka i razvitie voennogo dela* (Moscow: Voenizdat, 1976), pp. 109–11; I. I. Anureev, "Nauchno tekhnicheskii progress i voennaia nauka," *VM*, no. 2 (1970), pp. 27–39; M. Cherednichenko, "Nauchno tekhnicheskii progress i razvitie vooruzheniia i voennoi tekhniki," *VM*, no. 4 (1972), pp. 29–41. Edward Warner, the assistant secretary of defense for strategy and threat reduction, in June 1993, confirmed that the American definition of a Revolution in Military Affairs was heavily based on Russian or Soviet theoretical concepts; http://www.army.mil/cmh-pg/documents/Fletcher/Fletcher-99/F99-P5-T.htm.

67. Owens, *Lifting the Fog*, p. 83.

68. Soviet perspectives on the MTR were presented at the beginning of Krepinevich's assessment as working assumptions that provide a solid ground for developing further knowledge. Krepinevich, *Military-Technical Revolution*, pp. 6–8.

69. Marshall, *Some Thoughts on Military Revolutions*, pp. 3–6.

70. Allan R. Millett and Williamson Murray, eds., *Military Effectiveness*, vol. 1, *World War I*; vol. 2, *The Interwar Period*; and vol. 3, *The Second World War* (Cambridge: Cambridge UP, 1988); and Williamson Murray and Allan R. Millett, eds., *Military Innovation in the Interwar Period* (Cambridge: Cambridge UP, 1996).

71. Williamson Murray, ed., *National Security Challenges for the 21st century* (Carlislle, PA: US Army War College, 2003), p. 16.

72. Debra O. Maddrell, "Quiet Transformation: The Role of the Office of Net Assessment," The National Security Strategy Process, research paper (Washington, DC: National Defense University, 2003).

73. Maddrell, *Quiet Transformation*.

74. Krepinevich, *Military Technical Revolution*.

75. Tomes, *Military Innovation*, pp. 9–10; Der Derian, *Virtuous War*, pp. 28–29.

76. Bacevich, *New American Militarism*, pp. 164–66.

77. William Odom, "Soviet Force Posture: Dilemmas and Directions," *Problems of Communism* (July–August 1985), is probably the most precise open-source account on the Soviet MTR published not ex post facto, but simultaneously with the occurrence of the events.

78. Mahnken, *Technology*, p. 176.

79. Abella, *Soldiers of Reason*, pp. 279–80.

80. Maddrell, *Quiet Transformation*; Congressional Research Service report to the Congress, Theodor W. Galdi, *Revolution in Military Affairs?* (CRS 951170F, 11 December 1995); James Jay Carafano, "Dynamics of Military Revolution, 1300–2050," *Richmond Independent News*, 13 September 2002.

81. George Pickett, James Roche, and Barry Watts, "Net Assessment: A Historical Review," in *On Not Confusing Ourselves*, ed. Andrew Marshall, J. J. Martin, and Henry S. Rowen (Boulder, CO: Westview Press, 1991), pp. 158–85; Ken Silverstein, "The Man from ONA," *Nation*, 7 October 1999; Ken Silverstein and Daniel Burton-Rose, *Private Warriors* (New York: Verso, 2000), pp. 9–22; Fred Kaplan and Martin Sherwin, *The Wizards of Armageddon* (Stanford, CA: Stanford UP, 1991); Bill Gertz and Rowan Scarborough, "Yoda and the Jedis," *Inside the Ring*, 6 April 2001; Paul Bracken, "Net Assessment: A Practical Guide," *Parameters* 6 (Spring 2006); James Jay Carafano, "Stopping Surprise Attacks: Thinking Smarter about Homeland Security," *Heritage Foundation Backgrounder*, 23 April 2007; James Jay Carafano, "Dynamics of Military Revolution, 1300–2050," *Richmond Independent News*, 13 September 2002; Tom Ehrhard, "Net Assessment," syllabus, Strategic Studies Course 660.756, SAIS, Johns Hopkins University, 2007; Jason Vest, "A New Marshall Plan," *In These Times*, 2 April 2001; Jay Winik, "Secret Weapon," *Washingtonian* 34 (April 1999); Thomas Ricks, "Pentagon Study May Bring Big Shake-Up," *Washington Post*, 9 February 2001; John T. Bennett, "Pentagon Net Assessment Director Gets Medal," *Defense News*, 12 December 2008.

82. Abella, *Soldiers of Reason*, p. 277.

83. Bracken, "Net Assessment"; officially, net assessment was defined as "the comparative analysis of military, technological, political, economic, and other factors governing the relative military capability of nations. . . . [in order] to identify emerging or future threats and opportunities that deserve the attention of senior defense officials." Department of Defense Directive No. 5111.11, *Director of Net Assessment*, 21 August 2001.

84. Eliot Cohen, "Net Assessment: An American Approach," Jaffe Center for Strategic Studies Memorandum, no. 29 (Tel Aviv University, April 1990); The ONA and IDA Report, *Net Assessment: The Concept, Its Development and Its Future*, 22 May 1990.

85. Carafano, "Stopping Surprise Attacks"; Bracken, "Net Assessment"; Pickett, Roche, and Watts, "Net Assessment"; Jeffrey McKitrick, "Adding to Net Assessment," *Parameters* (Summer 2006), p. 119.

86. Rosen, "Net Assessment as an Analytical Concept," p. 295.

87. Bracken, "Net Assessment."

88. Pickett, Roche, and Watts, "Net Assessment," pp. 168–73.

89. Bracken, "Net Assessment"; McKitrick, "Adding to Net Assessment."

90. Bracken, "Net Assessment."

91. "Net assessments provided Secretaries of Defense with dynamic analysis of the way in which military conditions have changed in the past and the directions in which it may move in future." Rosen, "Net Assessment as an Analytical Concept," pp. 284, 299; Andrew Marshall, *Problems of Estimating Military Power* (Santa Monica, CA: RAND, 1966), "A Program to Improve Analytical Methods Related to Strategic Forces," *Policy Sciences* 15, no. 1 (November 1982), and "Arms Competitions: The Status of Analysis," in Uwe Nerlich, *The Soviet Asset: The Military Competition over Europe* (Cambridge, MA: Ballinger, 1983).

92. They concentrated on six fields: (1) measurement and forecast of trends in various military balances; (2) comparisons of weapons and force effectiveness; (3) historical evaluations of past performance of weapons used in specific conflict; (4) analysis of the role of perceptions of foreign decision makers and the process by which foreign institutions make strategic assessments; (5) search for new analytical and diagnostic tools; (6) analysis of particular issues of concern to the secretary of defense, highlighting important long-term trends, identifying future opportunities and risks in military competition, and appraising the strengths and weaknesses of US forces in light of long-term shifts in the security environment. Michael Pillsbury, *China Debates the Future Security Environment* (National Defense UP, 2000), appendix 1.

93. Owens, *Lifting the Fog*, pp. 80–81.

94. Andrew W. Marshall, "A Program to Improve Analytic Methods Related to Strategic Forces," *Policy Sciences* (November 1982).

95. Correspondence with Barry Watts, October–November 2006; conversation with Stephen Rosen, November 2006; conversation with Mary FitzGerald, October–November 2006; for example, see the accounts of Igor Birman, *Ia ekonomist: o sebe liubimom* (Moscow: Izdatelstvo Vremia, 2001).

96. Pickett, Roche, and Watts, "Net Assessment," pp. 174–75. Contributions of Notra Truelock and Mary FitzGerald can be seen as good examples.

97. Owens, *Lifting the Fog*, pp. 83–84.

98. Thomas Mahnken, *Technology and the American Way of War* (New York: Columbia UP, 2008), pp. 74–75; Barry Watts, *What Is the Revolution in Military Affairs?* (Northrop Grumman Analysis Center, 6 April 1995), pp. 1–2; correspondence with Barry Watts, October–November 2006; conversation with Stephen Rosen, November 2006; conversation with Mary Fitzgerald, October–November 2006.

99. Owens, *Lifting the Fog*, p. 83; Watts, *What Is the RMA?* pp. 1–2.

100. Der Derian, *Virtuous War*, pp. 30–32. Stuart Whitehead, director of the US Army TRADOC, argued that emulating holistic, synthetic, and nonlinear thinking from the Soviets and incorporating it with what is "tried and true" could change the entire US military culture, from training and education, to doctrine and equipment, to interagency and multinational cooperation. Whitehead in Murray, *Emerging Strategic Environment*, p. 49.

101. McKitrick, "Adding to Net Assessment," p. 119.

102. Maddrell, *Quiet Transformation*, pp. 3–6; Peter Boyer, "How Donald Rumsfeld Reformed the Army and Lost Iraq," *New Yorker*, 13 November 2006.

103. Galdi, *Revolution in Military Affairs?* p. 2.

104. Tomes, *Military Innovation*, p. 314.

105. An example of the former would be the RMA task forces established by Secretary of Defense William Perry in 1994. These task forces were to examine new warfighting concepts and related issues in the following areas: combined arms maneuver in theater warfare (later termed *dominant maneuver*), deep precision strike, forward operations, smaller-scale operations (special operations and peacekeeping), and fostering innovation. Watts, *What Is the RMA?* pp. 3–4.

106. "The intent was informing post–Cold War defense modernization activities with knowledge about military revolutions and the emerging technologies most likely to sustain a US military advantage." Tomes, *Military Innovation*, pp. 9, 128–41.

107. The ONA supported a number of historically based studies on military innovations during this and later periods. Studies sponsored by ONA in the 1990s were motivated by a broader question: "Looking back over the military history of the twentieth century, what were the fundamental technological, conceptual, operational, and organizational factors that, during times of peace, gave rise to fundamental changes in how military organizations would fight future wars?" Tomes, *Military Innovation*, p. 24; Barry Watts and Williamson Murray, "Military Innovation in Peacetime," in

Williamson Murray and Allan Millett, *Military Innovation in the Interwar Period* (New York: Cambridge UP, 1996). From 1993 to 1995 each of the services participated in RMA roundtables and war games financed and run by the ONA. Galdi, *Revolution in Military Affairs?* p. 9.

108. Steven Metz and James Kievit, *Strategy and the Revolution in Military Affairs: From Theory to Policy* (Carlisle, PA: US Army War College, 1995).

109. Mahnken, *Technology*, p. 177.

110. Vickers and Martinage, *Revolution in War*, p. i.

111. Colin Gray, *Out of the Wilderness: Prime Time for Strategic Culture* (Fort Belvoir, VA: Defense Threat Reduction Agency, 2006), p. 5; Vernon Loeb, "Billions, and It Can't Make Change," *Washington Post*, 13 September 2002.

112. Watts, *Six Decades*, p. 77; Watts, *What Is the RMA?* p. 6; Galdi, *Revolution in Military Affairs?* p. 9.

113. Watts, *What Is the RMA?* p. 6; Galdi, *Revolution in Military Affairs?* p. 9.

114. Colin Gray, *Recognizing and Understanding Revolutionary Change in Warfare* (Carlisle, PA: US Army War College, 2006), p. 8.

115. Richard Lock-Pullan, "The US Way of War and the War on Terror," *Politics and Policy* 34, no. 2 (2006), pp. 374–99.

116. Barry Watts, "Evolving Military Affairs," *Defense News*, 22 May 2006.

117. H. R. McMaster, "On War: Lessons to Be Learned," *Survival* 50, no. 1 (February–March 2008), pp. 19–30.

118. Edward Hall and Mildred Reed Hall, *Understanding Cultural Differences: Germans, French, and Americans* (Yarmouth, ME: Intercultural Press, 1990); Edward T. Hall, *Beyond Culture* (New York: Anchor Press, 1976); Geert Hofstede, *Culture's Consequences: International Differences in Work-Related Values* (Beverly Hills, CA: Sage, 1980); Charles Cogan, *French Negotiating Behavior: Dealing with La Grande Nation* (Washington, DC: United States Institute of Peace, 2003); Richard Solomon, *Chinese Negotiating Behavior: Pursuing Interest through "Old Friends"* (Washington, DC: United States Institute of Peace, 1999); and Robert Bathurst, *Intelligence and the Mirror: On Creating the Enemy* (New York: Sage Publications, 1993), describe the low-context American strategic behavior and a strong inclination to monochronism. Also see: Kevin Avruch, *Culture and Conflict Resolution* (Washington, DC: United States Institute of Peace Press, 1998), pp. 63–65; Raymond Cohen, *Negotiating across Cultures: International Communication in an Independent World* (Washington, DC: United States Institute of Peace Press, 1997), pp. 30–33.

119. Edward Stewart and Milton Bennett, *American Cultural Patterns: A Cross Cultural Perspective* (Yarmouth, ME: Intercultural Press, 1991), pp. 29–33; Cogan, *French Negotiating Behavior*; Gilbert Ryle, *The Aspects of Mind* (Cambridge, MA: Blackwell, 1993); Edmund Glenn and Christine Glenn, *Man and Mankind: Conflict and Communication between Cultures* (Norwood, NJ: Ablex, 1981), pp. 80–81, and

"Discussion and Debate on Conceptual Styles," *American Anthropologist* 71, no. 6 (December 1970), pp. 1448–50; Rosalie Cohen, "Conceptual Styles, Culture Conflict and Nonverbal Tests of Intelligence," *American Anthropologist* 71, no. 5 (October 1969), pp. 828–56.

120. Steven Sloman, "Feature Based Induction," *Cognitive Psychology* 25 (1993), pp. 231–80, and "The Empirical Case for Two Systems of Reasoning," *Psychological Bulletin* 119 (1996), pp. 30–52; Incheol Choi, Richard Nisbett, Ara Norenzayan, and Kaiping Peng, "Culture and Systems of Thought: Holistic vs. Analytic Cognition," *Psychological Review* 108 (2001), pp. 291–310; Herman Witkin, "Field Dependence and Interpersonal Behavior," *Psychological Bulletin* 84 (1977), pp. 661–89.

121. Stewart and Bennett, *American Cultural Patterns*, pp. 29–31.

122. Ryle, *Aspects of Mind;* Stewart and Bennett, *American Cultural Patterns*, pp. 32–33; Glenn and Glenn, *Man and Mankind*, pp. 80–81.

123. Stewart and Bennett, *American Cultural Patterns*, pp. 32–37; Neil Postman, *Technopoly: The Surrender of Culture to Technology* (New York: Knopf, 1992); Manfred Stanley, "Dignity versus Survival? Reflections on the Moral Philosophy of Social Order," in Richard Brown and Stanford Lyman, *Structure, Consciousness, and History* (Cambridge: Cambridge UP, 1978), pp. 197–234; Hall and Hall, *Understanding Cultural Differences.*

124. Pickett, Roche, and Watts, "Net Assessment," pp. 162–63.

125. Stewart and Bennett, *American Cultural Patterns*, pp. 35–36; Bathurst, *Intelligence and the Mirror.*

126. Kent Lam, Roger Buehler, Cathy McFarland, Michael Ross, and Irene Cheung, "Cultural Differences in Affective Forecasting: The Role of Focalism," *Personality and Social Psychology Bulletin* 31, no. 9 (2005), pp. 1296–1309; Timothy Wilson, Thalia Wheatley, Jonathan Meyers, Daniel Gilbert, and Danny Axsom, "Focalism: A Source of Durability Bias in Affective Forecasting," *Journal of Personality and Social Psychology* 78, no. 5 (May 2000), pp. 821–36.

127. The analytical style separates subjective experience from the inductive process that leads to an objective reality. The holistic style of thinking rests heavily on experience and fails to separate the experiencing person from objective facts, figures, or concepts. Students with "holistic" inclinations are said by their instructors to confuse concepts with impressions gained through observation or experience. In their writing and thinking, these students tend to give equal value to personal experience, empirical fact, and concepts derived from persons in authority. Cohen, "Conceptual Styles," pp. 853–55; Stewart and Bennett, *American Cultural Patterns*, p. 43.

128. Jae Min Jung and James Kellaris, "Cross-National Differences in Proneness to Scarcity Effects: The Moderating Roles of Familiarity, Uncertainty Avoidance, and Need for Cognitive Closure," *Psychology and Marketing* 21, no. 9 (September 2004), pp. 739–53; Ann Francoise Rutkowski and Bartel Van De Walle, "A Comparative

Analysis of Two Job Markets: Cultural Dimensions and Prototypical Criteria for Multi-Criteria Decision Support in Electronic Markets," *Group Decision and Negotiation* 14 (2005), pp. 285–306; Cheol Park and Jong Kun Jun, "A Cross-Cultural Comparison of Buying Behavior," *International Marketing Review* 20, no. 5 (2003); Elizabeth Wurtz, "Intercultural Communication on Web Sites: A Cross-Cultural Analysis of Web Sites from High-Context Cultures and Low-Context Cultures," *Journal of Computer-Mediated Communication* 11 (2006), pp. 274–99; Anthony Faiola and Sorin MaTei, "Cultural Cognitive Style and Web Design: Beyond a Behavioral Inquiry into Computer-Mediated Communication," *Journal of Computer-Mediated Communication* 11 (2006), pp. 375–94; Chang-Hoan Cho and Hongsik Cheon, "Cross-Cultural Comparisons of Interactivity on Corporate Web Sites," *Journal of Advertising* 34, no. 2 (Summer 2005).

129. Andrew May, *The Sources of the U.S. Military Advantage* (McLean, VA: SAIC, 2002), pp. 12, 14, 18, 26, 28, 36–37, 40, 42–43.

130. Chris Donnelly, *Red Banner: The Soviet Military System in Peace and War* (London: Jane's Information Group, 1988), p. 201.

131. Stewart and Bennett, *American Cultural Patterns*, pp. 41–42; Glenn and Glenn, *Man and Mankind*; Cohen, "Conceptual Styles"; Glenn and Glenn, "Discussion and Debate"; Cogan, *French Negotiating Behavior*; Bathurst, *Intelligence and the Mirror*; Raymond F. Smith, *Negotiating with the Soviets* (Bloomington: Indiana UP, 1989), pp. 5–6; Colin S. Gray, *Irregular Enemies and the Essence of Strategy: Can the American Way of War Adapt?* (Carlisle, PA: Strategic Studies Institute, US Army War College, 2006); Avruch, *Culture and Conflict Resolution*; Raymond Cohen, *Negotiating across Cultures*.

132. Bathurst, *Intelligence and the Mirror*, pp. 18, 26–29, 118–21.

133. The common law system is inductive in its essence: one starts with the particulars, upon which principles are then built. The legal system of the United States rests on cases and their evolution. This style of legal argumentation is at variance with European civil law, which is based on a deductive system in which one starts with a general rule and then applies the particulars. Cogan, *French Negotiating Behavior*, pp. 11, 48–49, 124–25, 247.

134. John Condon and Fatih Yousef, *An Introduction to Intercultural Communication* (Indianapolis: Bobbs-Merrill, 1975); Fred Jandt, *An Introduction to Intercultural Communication* (Thousand Oaks, CA: Sage, 1998); Matthew Bonham and Victor Sergeev, "The Limited Test-Ban Agreement: Emergence of New Knowledge Structures in International Negotiation," *International Studies Quarterly*, 41 (June 1997), pp. 215–40.

135. Brice F. Harris, *America, Technology and Strategic Culture: A Clausewitzian Assessment* (London: T&F Books, 2008), p. 88.

136. Colin Gray, *Strategy and History: Essays on Theory and Practice* (London: Routledge, 2006), p. 141; Ira Gruber, "The Anglo-American Military Tradition and the War for American Independence," in Kenneth J. Hagan and William R. Roberts,

Against All Enemies (Westport, CT: Greenwood Press, 1986), pp. 21–46; Ray Allen Billington, *America's Frontier Heritage* (Albuquerque: U of New Mexico P, 1986).

137. Allan R. Millett, "The United States Armed Forces in the Second World War," in Millett and Williamson Murray, eds., *Military Effectiveness*, vol. 3, *The Second World War* (Boston: Allen & Unwin, 1988), pp. 60–62, 81–84.

138. John Shy, *A People Numerous and Armed: Reflections on the Military Struggle for American Independence* (Ann Arbor: U of Michigan P, 1990), pp. 274–75.

139. May, *Sources of the U.S. Military Advantage*, pp. 23–24, 60.

140. Anita Arms, "Strategic Culture and the American Mind," in Thomas C. Gill, *Essays on Strategy IX* (Washington, DC: National Defense University, 1993), p. 25; Max Lerner, *America as a Civilization* (New York: Simon and Schuster, 1957), pp. 910–11; Colin Gray, *War, Peace and Victory: Strategy and Statecraft for the Next Century* (New York: Simon and Schuster, 1990), p. 354; Gray, "Strategy in the Nuclear Age," in Murray, Bernstein, and Knox, *Making of Strategy*, pp. 595–96.

141. May, *Sources of the U.S. Military Advantage*, pp. 7–8, 16.

142. Lock-Pullan, *US Intervention*, pp. 20–21; Russel F. Weigley, *Eisenhower's Lieutenants: The Campaign of France and Germany 1944–1945* (Bloomington: Indiana UP, 1990), p. 6.

143. Eliot Cohen, "The Strategy of Innocence? The United States, 1920–1945," in Murray, Bernstein, and Knox, *Making of Strategy*, p. 464. Also see: Adrian Lewis, *The American Culture of War* (London: Routledge, 2006), pp. 37–67.

144. John Ellis, *Brute Force: Allied Strategy and Tactics in the Second World War* (New York: Viking, 1990), pp. 534–35, xviii; Lock-Pullan, *US Intervention*, p. 18.

145. Martin Gannon, *Understanding Global Cultures* (Thousand Oaks, CA: Sage, 1994), p. 190; Gray, *Irregular Enemies*, pp. 35–36, 45–46; Thomas M. Kane, *Military Logistics and Strategic Performance* (London: Frank Cass, 2001); Gray, "Strategy in the Nuclear Age," p. 590.

146. Thomas Mahnken, *United States Strategic Culture* (Defense Threat Reduction Agency: SAIC, 2006), p. 10; also see Colin Gray, "Strategy in the Nuclear Age: The United States, 1945–1991," in Williamson Murray, Alvin Bernstein, and MacGregor Knox, *The Making of Strategy: Rulers, States, and War* (Cambridge: Cambridge UP, 1996), pp. 594–95.

147. Chester Wilmot, *The Struggle for Europe* (London: Wm. Collins, 1954), pp. 136–37; Millett, "United States Armed Forces in the Second World War," pp. 60–62, 81–84; Allan R. Millett and Peter Maslowski, *For the Common Defense: A Military History of the United States of America* (New York: Free Press, 1984), chs. 1–2.

148. Lock-Pullan, *US Intervention*, p. 23.

149. Richard Overy, *Why the Allies Won* (New York: Norton, 1995), p. 192, cited in Mahnken, *United States Strategic Culture*, p. 11; Bacevich, *New American Militarism*, pp. 156–58; Bathurst, *Intelligence and the Mirror*, p. 109. Albert Wohlstetter states that

superior economic resources offer numerous advantages in a war of attrition against materially inferior enemies. Albert Wohlstetter and Henry Rowen, *Objectives of the United States Military Posture* (RAND, 1 May 1959).

150. Brian M. Linn, "The American Way of War Revised," *Journal of Military History* 66, no. 2 (April 2002), pp. 501–33; Max Boot, *The Savage Wars of Peace: Small Wars and the Rise of American Power* (New York: Basic Books, 2002).

151. Mahnken, *United States Strategic Culture*, p. 10.

152. Ibid.; Matthew J. Morgan, "An Evolving View of Warfare: War and Peace and the American Military Profession," *Small Wars and Insurgencies* 16, no. 2 (June 2005), pp. 147–69; Jeremy Black, *Rethinking Military History* (London: Routledge, 2004).

153. Gray, *Irregular Enemies*, pp. 31, 38–39.

154. Eliot Cohen, "The Strategy of Innocence? The United States, 1920–1945," in Williamson Murray, Bernstein, and Knox, *The Making of Strategy: Rulers, States, and War* (Cambridge: Cambridge UP, 1996); Lock-Pullan, *US Intervention*; Ellis, *Brute Force*; Donn Starry, "A Perspective on American Military Thought," *Military Review* 69 (1989), pp. 2–11.

155. Loch-Pullan, *US Intervention*, pp. 19-22, 83; Gray, *Irregular Warfare*, p. 42.

156. Starry, "A Perspective on American Military Thought," pp. 2–11; Mahnken, *United States Strategic Culture*, p. 11.

157. May, *Sources of the U.S. Military Advantage*, p. 48; Lock-Pullan, *US Intervention*, p. 13; Gray, *Irregular Enemies*; Starry, "A Perspective on American Military Thought."

158. Lock-Pullan, *US Intervention*, p. 13; Williamson Murray, "The Future of American Military Culture: Does Military Culture Matter?" *Orbis* (Winter 1994), pp. 34–35.

159. Donnelly, *Red Banner*, p. 201.

160. Murray, "Future of American Military Culture?" pp. 34–35, and "Transformation of Professional Military Education," in *National Security Challenges for the 21st Century* (Carlisle, PA: US Army War College, 2003), pp. 10–11.

161. Colin Gray, "National Style in Strategy: The American Example," *International Security* 6, no. 2 (Fall 1981), pp. 35–37.

162. Murray, *Emerging Strategic Environment*, pp. 13–17.

163. Lacking any formal theory to approach the study of war, Western military researchers often had difficulty even grasping the terminology used by the Russians. Donnelly, *Red Banner*, pp. 101–2.

164. Ibid., p. 109.

165. Edward N. Luttwak, "The Operational Level of War," *International Security* 5, no. 3 (Winter 1980/81).

166. Raymond Garthoff, *Significant Features of Soviet Military Doctrine* (Santa Monica, CA: RAND, 1954); Erickson, "Significance of Operational Art."

167. Naveh, *In Pursuit*; F. F. Gaivoronovskii and M. I. Galkin, *Kultura voennogo myshlenia* (The Culture of Military Thought) (Moscow: Voenizdat, 1991).

168. Lock-Pullan, *US Intervention*, p. 102; Richard Simpkin, *Race to the Swift* (London: Batsford, 1985). The notion of operational art as a theoretical concept was rejected by the US military tradition until the 1980s. It was dismissed as a conceptually irrelevant notion with no clear aim or function, and as one that did not allow for the development of Western military art. Walter Jacobs, "Operational Art," *Army*, no. 11 (November 1961), p. 61, cited in Erickson, "Significance of Operational Art."

169. John Kiszely, "Thinking about the Operational Level," *RUSI* 150, no. 6 (December 2005), pp. 38–43; Richard Simpkin, *Deep Battle: The Brainchild of Marshall Tukhachevskii* (New York: Brassey's Defense Publishers, 1986), p. ix and ch. 5.

170. Lock-Pullan, *US Intervention*, p. 92.

171. Harris, *America*, pp. 120–21.

172. Max Weber, *The Protestant Ethic and the Spirit of Capitalism* (Gloucester, MA: Talcott Parsons, 1988), pp. 62–64; Barry Alan Shain, *The Myth of American Individualism: The Protestant Origins of American Political Thought* (Princeton, NJ: Princeton UP, 1995); Mark Noll, ed., *Religion and American Politics from the Colonial Period to the 1980s* (Oxford: Oxford UP, 1990), pp. 19–20; Philip Gleason, "American Identity and Americanization," *Harvard Encyclopedia of American Ethnic Groups* (Cambridge, MA: Harvard UP, 1980), pp. 31–32; Arms, "Strategic Culture and the American Mind," pp. 4–6.

173. May, *Sources of the U.S. Military Advantage*, pp. 12, 14, 18, 26, 28, 36–37, 40, 42–43, 45–46.

174. Gray, "Strategy in the Nuclear Age," p. 590, and *Irregular Enemies*, p. 33.

175. Stanley Hoffman, *Gulliver's Troubles: On the Setting of American Foreign Policy* (New York: McGraw-Hill, 1968).

176. Shy, *People Numerous*, pp. 238–40, 270; Charles Heller and William Stofft, eds., *America's First Battles, 1776–1965* (Lawrence, KS: Kansas UP, 1986); Gray, "Strategy in the Nuclear Age," p. 597, and *Irregular Enemies*, p. 33.

177. Gray, "Strategy in the Nuclear Age," pp. 588, 593.

178. Gray, "Strategy in the Nuclear Age," p. 598.

179. C. Vann Woodward, "The Age of Reinterpretation," *American Historical Review* 66 (October 1960).

180. Gray, "Strategy in the Nuclear Age," p. 598; Colin S. Gray, *Weapons Don't Make War: Policy, Strategy, and Military Technology* (Lawrence, KS: Kansas UP, 1993); and Gray, "Strategy in the Nuclear Age," pp. 608–9.

181. Bathurst, *Intelligence and the Mirror*, p. 120; Mahnken, *Technology*, p. 6.

182. Stuart Kinross, *Clausewitz and America: Strategic Thought and Practice from Vietnam to Iraq* (London: Routledge, 2008), pp. 2, 23–49; John Shy, "Jomini," in Peter Paret, *Makers of Modern Strategy: From Machiavelli to the Nuclear Age* (Princeton, NJ:

Princeton UP, 1986), pp. 149–50, 182–85; Gray, "Strategy in the Nuclear Age," pp. 588, 592–93; Lock-Pullan, *US Intervention*; Starry, "American Military Thought"; Philip Skuta, *Poker, Blackjack, Rummy and War: The Face of American Strategic Culture* (Carlisle, PA: US Army War College, 2006), pp. 10–11, 14–15; Carnes Lord, "American Strategic Culture," *Comparative Strategy* 5, no. 3 (1985), pp. 289–90.

183. Earl Tilford, *The Revolutions in Military Affairs: Problems and Cautions* (Strategic Studies Institute, US Army War College, 1995); Gray, "Strategy in the Nuclear Age," p. 598.

184. Arms, "Strategic Culture and the American Mind," pp. 9–12. Also see: Alvin Toffler, *Future Shock* (New York: Bantam Books, 1971), pp. 2–3, 54–55; David A. Hollinger, "The Problem of Pragmatism in American History," *Journal of American History* 67, no. 1 (1980), pp. 88–107. Americans do not resort to force quickly, but when they do, they expect it to lead to a victorious resolution. There is little patience for lengthy investments with distant payoffs; warfare is perceived as a regrettable and occasional evil that has to be settled as decisively and as rapidly as possible. Swift resolutions are frequently preferred over long-term and demanding enterprises. American sea and air doctrines argue that effective military action must be total. Walter Russell Mead, "The Jacksonian Tradition and American Foreign Policy," *National Interest* 58 (Winter 1999/2000), pp. 5–29; Gray, *War, Peace and Victory*, pp. 196, 354; Arms, "Strategic Culture and the American Mind," pp. 18–19; Frederick Downey and Steven Metz, "The American Political Culture and Strategic Planning," *Parameters*, September 1988, pp. 34–42; Cogan, *French Negotiating Behavior*, pp. 149–51; Shy, *People Numerous*, pp. 290–92.

185. Edward Hall, *The Dance of Life: The Other Dimensions of Time* (New York: Anchor Press, 1983), pp. 201–2, 221–23.

186. Stewart and Bennett, *American Cultural Patterns*, pp. 35–36; Bathurst, *Intelligence and the Mirror*, ch. 6.

187. Donnelly, *Red Banner*, pp. 123, 133.

188. Downey and Metz, "American Political Culture," pp. 34–42; Arms, "Strategic Culture and the American Mind," pp. 18–21.

189. For example, consider: post–Civil War reconstruction; the Marshall Plan; the leading US role in the UN and Breton Woods; the NSC-68; and the US commitment to containment of the USSR. Skuta, *Poker, Blackjack, Rummy and War*, pp. 16–17.

190. Jeremy Rifkin, *Time Wars: The Primary Conflict in Human History* (New York: Touchstone, 1987), pp. 73–74; Arms, "Strategic Culture and the American Mind," pp. 11–15; Skuta, *Poker, Blackjack, Rummy and War*, p. 7.

191. Murray, *Emerging Strategic Environment*, p. 16. This seems to signal a deep cultural difference between East and West. Donnelly notes that "Eastern émigrés to the West frequently complain about Western short-sightedness and short-term approach to problems." Donnelly, *Red Banner*, p. 31.

192. Murray, *Emerging Strategic Environment*, p. 16.

193. Eliot A. Cohen, "How to Think about Defense," in Williamson Murray, *1995–1996 Brassey's Mershon American Defense Annual* (Washington, DC: Brassey's, 1995).

194. Lawrence Korb, *The Joint Chiefs of Staff: The First Twenty-Five Years* (Bloomington: Indiana UP, 1976), and *The Fall and the Rise of the Pentagon* (Westport, CT: Greenwood Press, 1979); Thomas L. McNaugher and Roger L. Sperry, "Improving Military Coordination: The Goldwater-Nichols Reorganization of the Department of Defense," in Robert S. Gilmour and Alexis A. Halley, *Who Makes Public Policy?* (Chatham, NJ: Chatham House, 1994); Peter J. Roman and David W. Tarr, "The Joint Chiefs of Staff: From Service Parochialism to Jointness," *Political Science Quarterly* 113, no. 1 (1998), pp. 91–111; William J. Lynn and Barry R. Posen, "The Case for JCS Reform," *International Security* 10, no. 3 (Winter 1985–86), pp. 69–97.

195. Carl H. Builder, *The Masks of War: American Military Styles in Strategy and Analysis* (Baltimore: Johns Hopkins UP, 1989).

196. Joint Staff officers usually served only a single tour and regarded it as an unattractive career option—a dead end that provided few opportunities for promotion. As a result, it tended to be populated by the "soon-to-be retired." Ibid.; Roman and Tarr, "Joint Chiefs of Staff," p. 94.

197. Stephen Peter Rosen, *Winning the Next War: Innovation and the Modern Military* (Ithaca, NY: Cornell UP, 1994).

198. Roman and Tarr, "Joint Chiefs of Staff," pp. 91, 94; Lynn and Posen; Korb.

199. Evangelista; Bathurst, *Intelligence and the Mirror*. The US also approached negotiations in the same inductive, or bottom-up, manner. Cogan, *French Negotiating Behavior*, pp. 11, 48–49, 124–25, 247; Avruch, *Culture and Conflict Resolution*, pp. 63–65; Cohen, *Negotiating across Cultures*, pp. 30–33.

200. Murray, "Future of American Military Culture," pp. 39–41.

201. Some of them even recommended restructuring the General Staff system along the lines of the USSR or Germany. However, before the 1986 Goldwater-Nichols reform, little changed. Edward Luttwak, *The Pentagon and the Art of War* (New York: Simon and Schuster, 1984); Harold Brown, *Thinking about National Security: Defense and Foreign Policy in a Dangerous World* (Boulder, CO: Westview Press, 1983).

202. Colin S. Gray, *Irregular Enemies and the Essence of Strategy: Can the American Way of War Adapt?* (Carlisle, PA: Strategic Studies Institute, US Army War College, 2006), p. 36, *Strategy and History*, p. 166, and *Modern Strategy* (Oxford: Oxford UP, 1999), p. 147.

203. Mahnken, *United States Strategic Culture*, p. 12, and *Technology*, p. 4.

204. Skuta, *Poker, Blackjack, Rummy and War*, p. 16.

205. Gray, *Strategy and History*, p. 141; Kathleen Thelen, *How Institutions Evolve: The Political Economy of Skills in Germany, Britain, the United States, and Japan* (Cambridge: Cambridge UP, 2004), pp. 280–81.

206. Denis W. Brogan, *The American Character* (New York: Vintage Books, 1956), p. 150; Billington, *America's Frontier Heritage*; Gray, *Strategy and History*, p. 141; Ira Gruber, "The Anglo-American Military Tradition and the War for American Independence," in Hagan and Roberts, eds., *Against All Enemies*, pp. 21–46.

207. May, *Sources of the U.S. Military Advantage*, pp. 12, 14, 18, 26, 28, 35–37, 40, 42–43; Steven M. Irwin, *Technology Policy and America's Future* (New York: St. Martin's Press, 1993); Thomas Parke Hughes, *Changing Attitudes toward American Technology* (New York: Harper and Row, 1975).

208. Michael Sherry, *The Rise of American Air Power: The Creation of Armageddon* (New Haven, CT: Yale UP, 1987), pp. 233–35, and *In the Shadow of War: The United States since 1930* (New Haven, CT: Yale UP, 1997), pp. 10–11.

209. Tilford, *Revolution in Military Affairs*, p. 11; Loren Baritz, *Backfire: Vietnam, the Myths That Made Us Fight, the Illusions That Helped Us Lose, the Legacy That Haunts Us Still* (New York: William Morrow, 1984), p. 32.

210. Russell Weigley, *The American Way of War: A History of United States Military Strategy and Policy* (Bloomington: Indiana UP, 1977), p. 416; Gray, *Irregular Enemies*, p. 33; Sherry, *In the Shadow of War*, pp. 38–39, and *Rise of American Air Power*, pp. 233–35; Theo Farrell, *The Sources of Military Change: Culture, Politics, Technology* (London: Lynne Rienner Publishers, 2002), p. 19.

211. Weigley, *American Way of War*, p. 416.

212. Gray, "Strategy in the Nuclear Age," pp. 593, 609; Lock-Pullan, *US Intervention*, ch. 1, pp. 13–28.

213. Shy, *People Numerous*, pp. 287–88; Gray, *Irregular Enemies*, p. 37.

214. Tilford, *Revolution in Military Affairs*, pp. 10–11; Gray, *Strategy and History*, p. 165.

215. Andrew Krepinevich, *The Army and Vietnam* (Baltimore: The Johns Hopkins UP, 1988), pp. 5–7; Gray, *Irregular Enemies*, pp. 47–48; Theo Farrell, "Strategic Culture and American Empire," *SAIS Review* 25, no. 2 (Summer–Fall 2005); Lerner, *America as a Civilization*, p. 910; Jeffrey Record, "Collapsed Countries, Casualty Dread, and the American Way of War," *Parameters* (Summer 2002), pp. 4–23.

216. Farrell, "Strategic Culture and American Empire"; Gray, *Irregular Enemies*, pp. 37; 47–48, Krepinevich, *Army and Vietnam*, pp. 5–7; Chris Gray, *Postmodern War* (New York: Guilford Press, 1997), pp. 29, 50, 137, 225, 248; Lock-Pullan, *US Intervention*, p. 21; Robert Scales, *Firepower in Limited War* (Novato, CA: Presidio, 1995), pp. 3–5, 10–30.

217. Gray, *Postmodern War*, p. 137.

218. Eliot A. Cohen, "The Mystique of US Air Power," *Foreign Affairs* 73, no. 1 (1994), pp. 109–24; Harvey Sapolsky, Eugene Gholz, and Caitlin Talmadge, *US Defense Politics* (London: Routledge, 2009), pp. 23–27; Benjamin S. Lambeth, *The Transformation of American Air Power* (Ithaca, NY: Cornell UP, 2000).

219. Builder, *Masks of War*; Brian Linn, *The Echo of Battle: The Army's Way of War* (Cambridge, MA: Harvard UP, 2007).

220. Mahnken, *United States Strategic Culture*, pp. 12, 16–18; Builder, *Masks of War*, pp. 3–4; Farrell, "Strategic Culture and American Empire"; Murray, "Future of American Military Culture," p. 36; Thomas Mahnken and James FitzSimonds, *The Limits of Transformation: Officer Attitudes toward the Revolution in Military Affairs*, Newport Paper no. 17 (Newport, RI: Naval War College Press, 2003), ch. 6.

221. Russell Weigley, *The American Way of War: A History of United States Military Strategy and Policy* (Bloomington: Indiana UP, 1977), p. 416; Downey and Metz, "American Political Culture"; Todd Zachary, *The Effect of American Strategic Culture on Implementing National Strategy* (Maxwell AFB, AL: Air UP, 2000), pp. 50–53; Gray in Murray, pp. 35–36, 45–46; Lord, "American Strategic Culture"; Colin Gray, "Strategy in the Nuclear Age," pp. 588, 593; Tilford, *Revolution in Military Affairs*; May, *Sources of the U.S. Military Advantage*, pp. 45–48.

222. Ryle, *Aspects of Mind*; Stewart and Bennett, *American Cultural Patterns*, pp. 32–33; Glenn and Glenn, *Man and Mankind*, pp. 80–81; Arms, "Strategic Culture and the American Mind."

223. Mahnken, *United States Strategic Culture*, p. 13.

224. Francis J. Bremer, *John Winthrop: America's Forgotten Founding Father* (Oxford: Oxford UP, 2003); Gray, *Irregular Enemies*, p. 34, and "Strategy in the Nuclear Age," p. 591.

225. Booth, *Strategy and Ethnocentrism*; Gray, *Irregular Enemies*; Mahnken, *Uncovering Ways of War*; Robert H. Scales, "Culture-Centric Warfare," *US Naval Institute Proceedings* 130, no. 10 (October 2004), pp. 32–36; Cogan, *French Negotiating Behavior*, pp. 4, 6, 11; Archie Roosevelt, *For Lust of Knowing: Memoirs of an Intelligence Officer* (Boston: Little Brown, 1988), pp. 440–41; Irving Kristol, "Defining Our National Interest," *National Interest* (Fall 1990), pp. 19–20.

226. Bremer, *John Winthrop*.

227. Henry Steele Commager, *The American Mind: An Interpretation of American Thought and Character since the 1880s* (New Haven, CT: Yale UP, 1950), pp. 430–31; Gray, *War, Peace and Victory*, pp. 25–26; Arms, "Strategic Culture and the American Mind," pp. 24–25; Shy, *People Numerous*, pp. 278–82.

228. Gray, *Irregular Enemies*, p. 34, and "Strategy in the Nuclear Age," pp. 582, 591; Scales, "Culture-Centric Warfare," pp. 32–36; Shy, *People Numerous*, p. 268.

229. Shy, *People Numerous*, pp. 278–80; Gray, "Strategy in the Nuclear Age," pp. 582, 591, 597.

230. Shy, *People Numerous*, pp. 278–82, 285–86.

231. Kenneth J. Hagan and William R. Roberts, eds., *Against All Enemies: Interpretations of American Military History from Colonial Times to the Present* (Westport, CT: Greenwood Press, 1986), pp. 210–47; Shy, *People Numerous*, pp. 193–224.

232. The sustaining myth of American national exceptionalism—the notion of a truly unique society—fostered a strategic-cultural arrogance. Gray, "Strategy in the Nuclear Age," pp. 593–95. Even the collapse of the Soviet Union was understood by some scholars as proof of the superiority of American ideas about good governance and enlightened economics. Francis Fukuyama, "The End of History," *National Interest* (Summer 1989), pp. 3–18.

233. Richards J. Heuer, *Physiology of Intelligence Analysis* (Washington, DC: Center for the Study of Intelligence, 1999), pp. 70–71.

234. Andrew Stuart, *Friction in U.S. Foreign Policy: Cultural Difficulties with the World* (Carlisle, PA: Strategic Studies Institute, US Army War College, 2006).

235. Mahnken has detected signs of mirror imaging among American intelligence officers monitoring developments in Japan and in Nazi Germany during World War II. Mahnken, *Uncovering Ways of War*, pp. 11, 50.

236. Bathurst, *Intelligence and the Mirror*.

237. Donnelly, *Red Banner*, pp. 131–32, 206.

238. Gray, *Out of the Wilderness*, p. 23.

239. Mahnken, *Technology*, p. 5.

240. Pickett, Roche, and Watts, "Net Assessment," pp. 173–77.

Chapter 4

1. Avner Yaniv, *Politics and Strategy in Israel* (Tel Aviv: Sifriat Poalim, 1994), pp. 294–96 (Hebrew).

2. Ibid.; Dov Tamari, "The Quality of the Military," *Ma'arachot* (January 1979), pp. 2–6 (Hebrew), "Thoughts on Tactics," *Ma'arachot* (May–June 1980), pp. 2–5 (Hebrew), and "The Yom Kippur War: The Terms, the Assessments and the Conclusions," *Ma'arachot* (October–November 1980), pp. 11–16; Azar Gat, "On the Crises of Maneuver," *Ma'arachot* (October 1980), p. 43; Shmuel Gordon, "Who Needs the Pyrrhic Victory?" *Ma'arachot* (February 1983), pp. 34–37; Raanan Gissin, "Israeli Defensive Strategy: Change or Continuity?" *Ma'arachot* (February 1982), pp. 2–8.

3. Yossi Hochbaum, "Strengthening the Combined Arms Crew," *Ma'arachot* (February 1980), pp. 2–5 (Hebrew); Asa Beits, "Rapid Deployment Combined Arms Force: Reality or Imagination?" *Ma'arachot* (February 1982), pp. 16–22; Yaniv, *Politics and Strategy*, p. 384; Col. Ya'acov, "Command and Control on the Battlefield Directions and Tendencies," *Ma'arachot* (May–June 1980), pp. 57–60 (Hebrew).

4. Saadia Amiel, "Defensive Technologies for Small States: A Perspective," in *Military Aspects of the Israeli-Arab Conflict*, ed. Louise Williams (Tel Aviv: Tel Aviv University Publishing Projects, 1975), pp. 16–22, and "Deterrence by Conventional Means," *Survival* (March–April 1978), pp. 58–62.

5. Azar Gat, "Weaponry, Doctrine and Organization," *Ma'arachot* (January 1980), pp. 50–52 (Hebrew), and "Munitions, Concepts of Operations and Basic Organiza-

tions," *Ma'arachot* (January 1981), pp. 50–52; Zeev Bonen, "Weapons Development of the 1980s," *Ma'arachot* (April 1977); Maj. Nitzan, "Sacred Cows in Light of the Future Battlefield," *Ma'arachot* (February 1980), pp. 15–21 (Hebrew), and "The Future Battlefield: Another Conception," *Ma'arachot* (January–February 1981), pp. 15–18, 46–49; Zvi Lanir, "The Qualitative Factor in the Arab–Israeli Arms Race of the 1980s," *Ma'arachot* (February 1983), pp. 26–33 (Hebrew); Col. Ya'acov, "Command and Control on the Battlefield: Directions and Tendencies," *Ma'arachot* (May–June 1980), pp. 57–60 (Hebrew); Azriel Lorber, "Combat UAV," *Ma'arachot* 273 (June 1980), pp. 13–15 (Hebrew), and "Long-Range Artillery Shells," *Ma'arachot* (December 1982), p. 42; Yehuda Baal-Shem, "Technologies, Communications, Processing and Dissemination," *Ma'arachot* (July 1982), pp. 62–69; Raanan Gisin, "Defensive Strategy for Israel: Change or Continuity?" *Ma'arachot* 282 (February 1982), pp. 2–8 (Hebrew); Oded Tira, "Artillery: Weapon of Destruction," *Ma'arachot* (September 1983), pp. 15–17 (Hebrew); Zeev Klien, "Technological Race and Preparation for the Future," *Ma'arachot* 281 (November 1981), pp. 31–46 (Hebrew).

6. Yaniv, *Politics and Strategy*, p. 297.

7. Zeev Bonen, *Rafael: From the Laboratory to the Battlefield* (Israel: NDD Media, 2003), pp. 148–50 (Hebrew); Shmuel Gordon, *The Last Order of Knights: Modern Air Strategy* (Tel Aviv: Ramot, 1998), pp. 77–78 (Hebrew); Yehuda Weinraub, "The Israeli Air Force and the Air Land Battle," *IDF Journal* (Summer 1986), pp. 22–30; Yaniv, *Politics and Strategy*, p. 295; Jeffrey Isaacson, Christopher Layne, and John Arquilla, *Predicting Military Innovation* (Santa Monica, CA: RAND, 1999), pp. 31–32.

8. Gordon, *Last Order of Knights*, p. 78; Rebecca Grant, "The Bekaa Valley War," *Air Force Magazine* 85, no. 6 (June 2002); Eliezer Cohen and Zvi Lavi, *The Sky Is Not the Limit: The Story of the Israeli Air Force* (Tel Aviv: Sifriat Ma'ariv, 1990), pp. 611–18; Eliot Cohen, Michael Eisenstadt, and Andrew Bacevich, *Knives, Tanks, and Missiles: Israel's Security Revolution* (Washington, DC: Washington Institute for Near East Policy, 1998), p. 94; Trevor Dupuy and Paul Martell, *Flawed Victory: The Arab–Israeli Conflict and the 1982 War in Lebanon* (Fairfax: Hero Books, 1986), ch. 15.

9. Michael Handel, "The Evolution of Israeli Strategy: The Psychology of Insecurity and the Quest for Absolute Security," in Williamson Murray, Alvin Bernstein, and MacGregor Knox, *The Making of Strategy: Rulers, States and War* (Cambridge: Cambridge UP, 1996), p. 551; Benjamin Lambeth, *Moscow's Lessons from the 1982 Lebanon War* (Santa Monica, CA: RAND, 1984).

10. David Rodman, "Israel's National Security Doctrine: An Appraisal of the Past and a Vision of the Future," *Israel Affairs* 9, no. 4 (2003), pp. 127–28.

11. Some attempts to distill broader implications were made. Michael Handel, "Quantity versus Quality in Building Military Infrastructure," *Ma'arachot* 286 (February 1983), pp. 8–25 (Hebrew); Gordon, "Who Needs the Pyrrhic Victory?"; Dov Tamari, "Offense or Defense: Do We Have a Choice?" *Ma'arachot* 287 (October 1983),

pp. 5–11 (Hebrew); Lanir, "Qualitative Factor," pp. 26–33; Weinraub, "IAF and the Air Land Battle," pp. 22–30.

12. Cohen, Eisenstadt, and Bacevich, *Knives, Tanks, and Missiles*, p. 82.

13. Yaniv, *Politics and Strategy*, p. 385.

14. Cohen, Eisenstadt, and Bacevich, *Knives, Tanks, and Missiles*, p. 77.

15. Isaac Ben-Israel, "Where Did Clausewitz Go Wrong?" *Ma'arachot* (February–March 1988), pp. 16–26 (Hebrew).

16. Yaniv, *Politics and Strategy*, pp. 389–90; Erez Winer, "Who Needs the Doctrine?" *Ma'arachot* 385 (September 2000), p. 43 (Hebrew).

17. Yaniv, *Politics and Strategy*, pp. 385–88.

18. Israel Tal, "The Offensive and the Defensive in Israel's Campaigns," *Jerusalem Quarterly* (Summer 1989); Moshe Bar-Kochba, "Security Concept and Readiness Aspects under the 1973 Test," *Ma'arachot* 316 (June 1989), pp. 7–11 (Hebrew).

19. Ariel Levite, *Offense and Defense in Israeli Military Doctrine* (Tel Aviv: Jaffe Center for Strategic Studies, 1988) (Hebrew); Shmuel Gordon, "Principles of Combat with PGMs," *Ma'arachot* (April 1987), pp. 22–42 (Hebrew); Bonen, *Rafael*, pp. 129–34; Reuven Pedatzur, "Israel: Update of Military Doctrine," *Ma'arachot* 319 (January 1990), pp. 20–29 (Hebrew); Yossi Peled, "Does Israeli Operational Concept Need a Change?" *Ma'arachot* 316 (January 1990), pp. 2–5 (Hebrew); Chris Belmi, "Artillery in the Modern Battlefield," *Ma'arachot* 318 (January 1990), pp. 16–31 (Hebrew); Yossi Hochbaum, "PGMs in World Militaries: Doctrines and New Conceptual Aspects," *Ma'arachot* 302 (April 1986), pp. 6–64 (Hebrew); Yehezkel Dror, "Strategic-Political Considerations for Updating Military Doctrine," *Ma'arachot* 310 (December 1987), pp. 8–12 (Hebrew); Azriel Lorber, "Should the Tank Disappear?" *Ma'arachot* 305 (September 1986), pp. 8–12 (Hebrew); Yossi Zinger, "Advanced Artillery," *Ma'arachot* 297 (January 1985), pp. 39–42 (Hebrew).

20. Yaniv, *Politics and Strategy*, pp. 389–90.

21. Stuart Cohen, "Israel's Changing Military Commitments, 1981–1991," *Journal of Strategic Studies*, no. 6 (1992), pp. 330–50, "The Peace Process and Its Impact on the Development of a Slimmer and Smarter IDF," *Israel Affairs* 1, no. 1 (Summer 1995), pp. 1–21, "Small States and Their Armies: Restructuring the Militia Framework of the IDF," *Journal of Strategic Studies* 18, no. 4 (December 1995), and "Towards a New Portrait of a New Israeli Soldier," *Israel Affairs* 3 (Spring/Summer 1997), pp. 77–117; Moti Geva, "The IDF Ethos," *Ma'arachot* (November 2000), pp. 25–31 (Hebrew).

22. Shimon Naveh, "Defense of Israel in the 21st Century," *Ma'arachot* 355 (January 1998), pp. 46–52 (Hebrew); Cohen, Eisenstadt, and Bacevich, *Knives, Tanks, and Missiles*, pp. xvi, 92; Gabriel Ben-Dor, "Responding to the Threat: The Dynamics of Arab–Israel Conflict," in Daniel Bar-Tal, Dan Jacobson, and Aharon Klieman, eds.,

Security Concerns: Insights from the Israeli Experience (Stamford, CT: JAI Press, 1998), pp. 130–33.

23. Zeev Bonen, "Sophisticated Conventional Warfare," in *Advanced Technology and Future Warfare*, ed. Eliot Cohen and Zeev Bonen (Ramat Gan: Bar Ilan University, 1997); Shmuel Gordon, "Technological Warrior and the War in Lebanon," *Ma'arachot* 350 (January 1997), pp. 20–29; Cohen, Eisenstadt, and Bacevich, *Knives, Tanks, and Missiles*, p. 99; Kenneth Werel, "The Winning Air Power: The Gulf War and Vietnam," *Ma'arachot* 330 (June 1993), pp. 18–27 (Hebrew); Reuven Pedatzur, "The Gulf War in the Eyes of Israeli Observers," *Ma'arachot* 330 (June 1993), pp. 5–17 (Hebrew); Israel Tal, "The Gulf War: Some Lessons," *Ma'arachot* 323 (March 1992), pp. 2–5 (Hebrew); Chris Demchak, "Technological Implications of the Gulf War," *Ma'arachot* 323 (March 1993), pp. 22–31 (Hebrew); Col. Y., "The Gulf War: The First Lessons for Israel," *Ma'arachot* 322 (November 1991), pp. 2–7 (Hebrew).

24. David Rodman, *Defense and Diplomacy in Israel's National Security Experience* (Portland, OR: Sussex Academic Press, 2005), p. 14, and "Israel's National Security Doctrine," p. 128.

25. Dan Shomron, "The IDF Perspective," *IDF Journal*, no. 17 (Summer 1989), pp. 3–5; Ehud Barak, "Interview," *IDF Journal*, no. 20 (Summer 1990), pp. 13–18; Shmuel Gordon, "In Favor of Selective Conscription," *Ma'arachot* (1993), pp. 32–37 (Hebrew); Cohen, Eisenstadt, and Bacevich, *Knives, Tanks, and Missiles*, p. 99; Aluf Ben, "The Peace Is Important, but the IDF Is More," *Ha'aretz*, 27 July 1999, and "Israeli Peace Forces," *Ha'aretz*, 19 November 1993.

26. Reuven Pedatzur, "Updating Israel's Military Doctrine," *IDF Journal* (Winter 1991), pp. 32–35; Handel, "The Evolution of Israeli Strategy," pp. 540, 557; Dov Tamari, "Does a Conservatism of Thought Exist in the IDF?" *Ma'arachot* (October–November 1989), pp. 23–35; Ariel Levite, ed., *Decision on the Modern Battlefield* (Hebrew) (Tel Aviv: Jaffe Center for Strategic Studies, 1990); Azriel Lorber, *Precision Guided Munitions in the Ground Forces' Battle* (Hebrew) (Tel Aviv: Ma'arachot, 1992); Zeev Bonen, "Technology in War: Preliminary Lessons from the Gulf War," in *War in the Gulf: Implications from Israel* (Tel Aviv: Jaffe Center for Strategic Studies, 1992), and "The Impact of Technological Developments on the Strategic Balance in the Middle East," in *The Middle East Military Balance: 1993–1994*, ed. Shlomo Gazit, (Boulder, CO: Westview, 1994), pp. 148–63; Cohen, Eisenstadt, and Bacevich, *Knives, Tanks, and Missiles*, p. xv; Jonathan Marcus, "Israel Sharpens Its Military Strategy," *BBC News Special Report* (29 April 1998).

27. Editorial, "The Range Extension," *Ma'arachot* 338 (November 1994), pp. 40–45 (Hebrew); Edward Pin, "New Technologies of the Ground Forces," *Ma'arachot* 331 (August 1993), pp. 10–19 (Hebrew); Isaac Ben-Israel, "About the Laws of Physics and Command and Control Systems," *Ma'arachot* 326 (August 1992), pp. 40–42 (Hebrew), "Technological Lessons," *Ma'arachot* 332 (October 1993), pp. 8–13 (Hebrew), and "Back to the

Future," *Ma'arachot* 329 (April 1993), pp. 2–45 (Hebrew); General Y., "Targets, Fire Support and Smart Munitions," *Ma'arachot* 328 (January 1993), pp. 2–8 (Hebrew).

28. Steven Silvesy, "Air-Land Battle Future: The Tactical Battlefield," *Ma'arachot* 322 (March 1991), pp. 8–15 (Hebrew); Shmuel Gordon, *The Bow of Paris: Technology, Doctrine and Israeli Security* (Tel Aviv: Sifriat Ha Poalim, 1997), see esp. pp. 5–6.

29. Uri Bar-Joseph, "Israel's Northern Eyes and Shield: The Strategic Value of the Golan Heights Revisited," *Journal of Strategic Studies* 21, no. 3 (September 1998), pp. 46–66; Ilan Amit, "Prudence and Technology on the Future Battlefield," *Ma'arachot* 297 (January 1985), pp. 35–38 (Hebrew); Azriel Lorber, "Future Joint Battlefield," *Ma'arachot* 319 (July 1990), pp. 40–43 (Hebrew), and "On the Technological Implications of the Gulf War," *Ma'arachot* 321 (June 1992), pp. 36–41 (Hebrew); Yehida Drori, "The Future Battlefield and the Social and Economic Aspects," *Ma'arachot* 334 (February 1994), pp. 55–57 (Hebrew); Amnon Sofer, "Future Battlefield in the Middle East," *Ma'arachot* 368 (December 1999), pp. 19–25 (Hebrew); Avi Kober, "War and Future Battlefield," in *Israel's Security Web*, ed. Haggai Golan (Tel Aviv: Ma'arachot, 2001) (in Hebrew), pp. 269–73.

30. Ed Blanche, "Israel Addresses the Threats of the New Millennium," *Jane's Intelligence Review* 11, no. 2 (February 1999), pp. 24–26, and 11, no. 3 (March 1999), pp. 27–28; Amir Rappoport, "The First Round of Discussions for Updating the Security Concept Has Ended," *Yediot Ahronot*, 30 June 1998 (Hebrew); Alex Fishman, "Preparing for the Last War," *Yediot Ahronot*, 4 December 1998 (Hebrew).

31. Cohen, Eisenstadt, and Bacevich, *Knives, Tanks, and Missiles*, pp. 1, 78, 91, 129; National Intelligence Counsel Conference Report Summary, *Buck Rogers or Rock Throwers?* 14 October 1999.

32. Dovik Tamari, "The System of Operational Concepts and the IDF Strategy," in *Revolution in Military Affairs: The Summary of the Israeli–American Meeting in October 1997*, Observation no. 20 (TOHAD, May 1998), p. 71 (Hebrew); Shimon Naveh, "The Cult of Offensive Preemption and Future Challenges for Israeli Operational Thought," in Efraim Karsh, *Between War and Peace: Dilemmas of Israeli Security* (London: Frank Cass, 1996), pp. 170–77, and "Was Blitzkrieg a Military Doctrine?" *Ma'arachot* 330 (May–June 1993), pp. 28–54, and 331 (July–August 1993), pp. 30–40, and *In Pursuit of Military Excellence: The Evolution of Operational Theory* (London: Frank Cass, 1997), pp. 105–38.

33. Tamari, "System of Operational Concepts," p. 76; Haggai Golan and Shaul Shay, *Low-Intensity Conflict* (Tel Aviv: Ma'arachot, 2004), p. 490; Operational Theory Research Institute PowerPoint Presentation, pp. 6–8; Shimon Naveh, *Encounter in Samarra: Texts on Strategic Engagement* (IDF: OTRI, 2002), pp. 2–8.

34. Cohen, Eisenstadt, and Bacevich, *Knives, Tanks, and Missiles*, p. 78; Operational Theory Research Institute, *Presentation to the GS Committee* (IDF: OTRI, September 2000), p. 1; Caroline Glick, "Halutz's Stalinist Moment," *Jerusalem Post*, 8 June 2006; Kobi

Michael, "The Israeli Defense Forces as an Epistemic Authority: An Intellectual Challenge in the Reality of the Israeli–Palestinian Conflict," *Journal of Strategic Studies* 30, no. 3 (June 2007), pp. 421–46; Yadidia Ya'ari, "Decentralized Warfare and Dynamic Molecule," *Ma'arachot* 395 (October 2004), p. 11, n. 1 (Hebrew); Shaul Bronfeld, "Trivial Conclusions or Hidden Agenda?" *Ma'arachot* 366–67 (October 1999), pp. 84–85 (Hebrew); Zvi Lanir, "Doctrines and Systems of Conceptualization and Interpretation," *Ma'arachot* 355 (January 1998), pp. 56–59 (Hebrew); Yedidia Ya'ari and Haim Assa, *Diffused Warfare: War in the 21st Century* (Tel Aviv: Yediot Aharonot, 2005), pp. 9–16 (Hebrew).

35. Tamari, "System of Operational Concepts," p. 76; Zvi Lanir, "From Operational Art to Systemic Thinking," *Ma'arachot* 352–53 (August 1997), pp. 2–19 (Hebrew); Shimon Naveh, "Operational Art and General Systems Theory," *Ma'arachot* 344–45 (December 1995–January 1996) (Hebrew).

36. For the most nuanced presentation of this quintessence, see: Shimon Naveh, "Asymmetrical Conflicts: Operational Critique of the Hegemonic Strategies," in Golan and Shay, *Low-Intensity Conflict*, pp. 101–45. Also, for reflections on the epistemology conference, see *Martial Ecologies*, organized by OTRI; Chris Gray, *Peace, War, and Computers* (London: Routledge, 2005), pp. 104–6.

37. Avi Kober, "The Intellectual and Modern Focus in Israeli Military Thinking as Reflected in *Ma'arachot* Articles, 1948–2000," *Armed Forces and Society* 30, no. 1 (Fall 2003), p. 155.

38. Naveh, *In Pursuit, Operational Art: The Emergence of Military Excellence* (Tel Aviv: Ma'arachot, 2001) (Hebrew), and "Operational Art and Systems Theory"; Lanir, "From Operational Art."

39. Mostly works by Mikhail Tukhachevsky, Georgy Isserson, Nikolai Varfolomeev, and Vladimir Triandafillov. Naveh, "Cult of Offensive Preemption," pp. 181, 187, nn. 46–50; Lanir, "From Operational Art," pp. 11–13.

40. Shimon Naveh, "New Elements of Maneuver: Soviet Revolutionary Operational Concept," *Ma'arachot* 324 (May 1992), pp. 20–29, and 326 (August–September 1992), pp. 12–23 (Hebrew), and "Operational Art and the General Systems Theory," p. 3; Lanir, "From Operational Art," p. 15.

41. For example, the language used to criticize the IDF's conduct during the 1982 war clearly emulates Soviet operational terminology. Naveh, "Cult of Offensive Preemption," p. 181. For comparison with the Soviet usage of the same terms in a similar pattern, see: P. G. Skachko, "Simultaneous Attack along the Entire Depth of the Operational Deployment of the Enemy—the Main Tendency in the Current Development of the Operational Art," *VM*, no. 7 (1985), pp. 18–24 (Russian).

42. Lanir, "From Operational Art," pp. 7, 13, 18; Naveh, "Operational Art and Systems Theory," p. 10.

43. Naveh, *In Pursuit*, pp. 210–36, 322–31, "Cult of Offensive Preemption," pp. 181–82, and "Operational Art and Systems Theory," pp. 10–11. *Fragmentation strike*

stands for the Soviet *udar*; *simultaneity* for the Soviet *sinkhronizatciia, soglasovannost'*, and *vzaimodeistvie*; *momentum* stands for the Soviet *temp operatcii* and *razvitie uspekha*. Lanir, "From Operational Art," pp. 2–13.

44. Zvi Lanir and Gad Sneh, *The New Agenda of Praxis* (Tel Aviv: Praxis, 2000), pp. 21–22; Naveh, "Operational Art and Systems Theory."

45. Lanir, "From Operational Art," pp. 7, 13, 18; Naveh, "Operational Art and Systems Theory," p. 10.

46. Naveh, "Cult of Offensive Preemption," p. 181, and "Operational Art and Systems Theory," p. 6; Lanir, "From Operational Art," esp. pp. 11–12.

47. Lanir, "From Operational Art," p. 18.

48. The first block included works by Mikhail Tukhachevsky, Georgy Isserson, Nikolai Varfolomeev, and Vladimir Triandafillov; the second block included works by Thomas Kuhn, Bernard Cohen, and M. H. F. Wilkins; the third block included Humberto Maturana, Stuart Umpleby, Georg von Krogh, and Johan Roos; the fourth block consisted of Paul Virilio, Gilles Deleuze, and Felix Guattari. Shimon Naveh and Amos Granit, *Operational Art as Field of Knowledge: Advanced Operational Course Textbook* (Tel Aviv: OTRI, 1998); Shimon Naveh, *Cybernetics and Design: Recommended Book List* (Tel Aviv: OTRI, 1999); *KOMEM no. 5 Syllabus: Knowledge Spaces—Strategic Discourse* (Tel Aviv: OTRI, 2001); James Der Derian, *Virtuous War* (New York: Westview Press, 2001), pp. 219–20; Eyal Weizman, "Lethal Theory," Roundtable: Research Architecture (9 May 2006), p. 67, http://roundtable.kein.org/node/415, and "Walking through Walls: Soldiers as Architects in the Israeli–Palestinian Conflict," *Radical Philosophy* 136 (March–April 2006), pp. 5–6.

49. Weizman, "Lethal Theory," pp. 50, 67, 69.

50. Michael, "Israeli Defense Forces," p. 432; Naveh, "Operational Art and Systems Theory"; "Presentations by Doron Almog and Zvi Lanir," in Dan Moore, *US–Israeli Seminar on Military Innovation and Experimentation, March 1999* (Joint Advanced War-Fighting Program, Institute for Defense Analysis, SAIS, 1–2 March 1999); also see Lanir and Sneh, *New Agenda of Praxis*, pp. 20–22; Zvi Lanir, "The Strategic Cognitive Threshold," *Ma'arachot* (August–September 1992), pp. 2–11 (Hebrew); Weizman, "Lethal Theory," p. 67, and "Walking through Walls," pp. 5–6; Izhar Sahar and Keren Zur, "Knowledge Development," *Ma'arachot* 368 (December 1999), pp. 62–67 (Hebrew).

51. Lanir and Sneh, *New Agenda of Praxis*; Zvi Lanir, "The Principles of War and Military Thinking," *Journal of Strategic Studies* 16, no. 1 (March 1993), *The Revolution in Military Affairs: An Israeli Perspective* (Tel Aviv: Praxis, 1996), *Military Systemic Thinking* (Tel Aviv: Praxis, 1996), "From Operational Art to Systemic Thinking," pp. 2–19, "The Failure of Military Thought in Low-Intensity Conflict," *Ma'arachot* 365 (September 1999), pp. 4–13, and "Military Doctrine and System of Interpreting Concepts," *Ma'arachot* 355 (January 1998), pp. 56–59 (Hebrew); Lanir and Almog, Presentations in Moore, US-Israeli Seminar.

52. Michael, "Israeli Defense Forces," p. 432; Lanir and Sneh, *New Agenda of Praxis*, pp. 20–24. For the application of the OTRI methodology, see: Shirli Karni, "Operation Yoav: Systemic Analysis," *Ma'arachot* 365 (August 1999), pp. 43–55 (Hebrew); Ophra Graitzer, "Deep Range Penetration: Wingate's Paths in Burma as Alternative Foundation for the Operational Space," *Ma'arachot* 392 (December 2003), pp. 42–47 (Hebrew).

53. Lanir, "From Operational Art," pp. 3–4, 7–8; Zeev Schiff, "The New Estimation of the Situation," *Ha'aretz*, 22 August 2000.

54. Tamari, "System of Operational Concepts," p. 72; Michael, "Israeli Defense Forces," p. 426, and "Military Knowledge as the Basis for Arranging the Discourse Space," *Ma'arachot* 403–4 (December 2005), pp. 44–46 (Hebrew); Naveh, "Operational Art and Systems," p. 9; Lanir and Sneh, *New Agenda of Praxis*, pp. 10, 25; Moshe Ya'alon, "The Challenge of Educating the IDF Officers," *Ma'arachot* 396 (September 2004), pp. 4–5 (Hebrew).

55. Michael, "Israeli Defense Forces," p. 434; Weizman, "Lethal Theory," p. 60.

56. Weizman, "Lethal Theory," pp. 53–67, and "Walking through Walls"; Nicolai Ouroussoff, "As Israeli Barrier Goes Up, Views Harden on All Sides," *International Herald Tribune*, 10 January 2006; Gershon Hacohen, "Military Professional and Operational Planning: From Engineering to Architecture," *Ma'arachot* 376 (April 2001), pp. 10–15 (Hebrew); Chen Kotes-Bar, "Starring Him: Interview with BG Aviv Kokhavi," *Ma'ariv*, 22 April 2005; Barbara Opall-Rome, "Fake, Flexible City Rises in the Negev: New Israeli Training Facility to Host U.S. Troops," *Defense News*, 11 June 2007, p. 8; Amir Golan, "The Components of the Ability to Fight in Urban Areas," *Ma'arachot* 384 (July 2002).

57. Michael, "Israeli Defense Forces," pp. 431–32. Classical military terms were replaced by surrogates from general systems theory. For example, *systemic tendency, commander of the operational tendency, operator, operational tension, operational lever, operational tow, burning and producing effects on the enemy consciousness, operational strike on the enemy system, sources of systemic tension, formless enemy systems*, and *deconstruction of the enemy system from its rationale*. Amir Rappoport, "Warning: Swarms with Teeth," *Ma'ariv* (13 October 2007), pp. 7–11.

58. Michael, "Israeli Defense Forces," pp. 433–34; Ya'ari and Assa, *Diffused Warfare*, pp. 9–16, esp. p. 11.

59. Michael, "Israeli Defense Forces," pp. 435–36; Amnon Barzilai, "Getting the Aftermath Right," *Ha'aretz*, 24 April 2004.

60. Interview with Ariel Levite, then senior official in the Israeli Ministry of Defense, Tel Aviv, 9 August 2006.

61. Andrew Marshall, "Introduction," pp. 2–4; Colonel Rowell, "What Is RMA?" pp. 4–10; Michael Vickers, "RMA in the US Army," pp. 10–21; James Blackwell, "Dominant Maneuver," pp. 21–53, all in *Revolution in Military Affairs: The Summary of the Israeli–American Meeting*.

62. Doron Almog, "The Discussion from Different Positions," pp. 66–71; Tamari, "System of Operational Concepts," pp. 71–85; Shimon Naveh, "Operational Engagement Meeting," pp. 85–96, all in *Revolution in Military Affairs: The Summary of the Israeli–American Meeting*.

63. *Revolution in Military Affairs: Summary of the Israeli–American Meeting*, pp. 66–69, 73–74; also see: Cohen, Eisenstadt, and Bacevich, *Knives, Tanks, and Missiles*, p. 1.

64. *Revolution in Military Affairs: Summary of the Israeli–American Meeting*, pp. A, 96.

65. For contacts and cooperation between OTRI and ONA since 1997, see: Moore, *US–Israeli Seminar*; Amir Oren, "IDF to Share Its Anti-Terrorism Expertise with Pentagon," *Ha'aretz*, 16 February 2005; Barzilai, "Getting the Aftermath Right," "IDF Prepared for Changed Warfare," *Ha'aretz*, 7 January 2004, and "The Tank of the Future," *Ha'aretz*, 17 October 2003; Barry Watts, *Six Decades of Guided Munitions and Battle Networks* (Washington, DC: CSBA, 2007), pp. 64–65; Tim Challans, "Emerging Doctrine and the Ethics of Warfare," presentation for the Joint Services Conference on Professional Ethics (Ft. Leavenworth, KS: SAMS, June 2006), n. 7.

66. "OTRI: Operational Theory Research Institute," PowerPoint presentation (OTRI, September 2004), slides 5, 14, 16; Editorial, "IDF Team to Brief Pentagon on Counterinsurgency Tactics," *Ha'aretz*, 15 October 2001; Barzilai, "Getting the Aftermath Right."

67. *Revolution in Military Affairs: Summary of the Israeli–American Meeting*, p. 96.

68. *The Military Technical Revolution*, Observation no. 15 (Tel Aviv: TOHAD, June 1995) (Hebrew); *The Russian Military Technical Revolution*, Observation no. 16, (Tel Aviv: TOHAD, November 1996) (Hebrew); *The US Air Land Battle* (Tel Aviv: IDF/TOHAD, July 1993) (Hebrew); *US Army Doctrine of Ground Operations: Full Dimensional Battlefield; FM 100–5, Operations* (Tel Aviv: IDF/TOHAD, 1994) (Hebrew).

69. *Aspects of Operational Art*, Observation no. 12 (Tel Aviv: TOHAD, November 1993) (Hebrew); Led Holder, "The Renaissance of Operational Art," *Ma'arachot* 345 (January 1996), pp. 48–60 (Hebrew); William Bolt and Daviv Yablonsky, "Tactics and Operational Level in War," *Ma'arachot* 345 (January 1996), pp. 36–45 (Hebrew); Major Dan, "The Development of the American Operational Doctrine (Air-Land Battle) and Its Characteristics," *Ma'arachot* 346 (February 1996), pp. 14–17 (Hebrew); Zeev Bonen, "The Road to the Three-Dimensional War," *Ma'arachot* 340 (February 1995), pp. 2–10 (Hebrew), and "The Role of the Air Force in the Modern Land Battle," *Ma'arachot* 346 (February 1996), pp. 2–5 (Hebrew).

70. Yaacov Zur and Beny Bet-Or, "The Demands of Intelligence in the Advanced Military," *Ma'arachot* 334 (February 1994), pp. 42–50 (Hebrew); Editorial, "Precision Guided Munitions," *Ma'arachot* 349 (July 1996), pp. 44–51 (Hebrew); Lt. C. Erez, "On Procuring New Technologies," *Ma'arachot* 348 (June 1996), pp. 20–25 (Hebrew); Uri Rom, "Destruction of Targets in the Speed of Light," *Ma'arachot* 346 (February 1996),

pp. 11–13 (Hebrew); Azriel Lorber, "On the Imagined Border," *Ma'arachot* 350 (September 1996), pp. 41–52.

71. Peter Ying, "Developments in Land Warfare," *Ma'arachot* 340 (February 1995), pp. 30–35 (Hebrew); Beny Michelson, "The US Military in the 21st Century," *Ma'arachot* 339 (February 1995), pp. 10–16 (Hebrew); Gordon Sullivan and James Dubick, "War in the Information Era," *Ma'arachot* 346 (February 1996), pp. 40–50, and "Ground Warfare in the Twenty-First Century," *Ma'arachot* 347 (April 1996), pp. 32–42 (Hebrew).

72. Azriel Lorber, "On the Next Military Revolution," pp. 13–20; Gal Loft, "Command and Leadership in the Era of Military Technical Revolution," pp. 60–64; Eli Shvili, "The Importance of Countermeasures," pp. 38–45; Eran Lerman, "Thoughts about Revolution in Military Affairs and Problematic Combination of Operational Art and Intelligence," pp. 21–37; Eliot Cohen, "American Vision of the Revolution in Military Affairs," pp. 2–12, all in *Ma'arachot* (January 1998) (Hebrew).

73. Eliot Cohen, Michael Eisenstadt, and Andrew Bacevich, "Revolution in Israeli Security Affairs," *Ma'arachot* 364 (June 1999), pp. 44–51 (Hebrew). Even the Soviet MTR appeared: Colonel Yair, "Active Measures against PGM Threat on the Tactical Battlefield—the Russian Approach," *Ma'arachot* 363 (March 1999), pp. 30–37.

74. Moshe Sukenik, "Do Lessons from Kosovo Diminish the Role of the Ground Maneuver?" *Ma'arachot* 368 (March 2000), pp. 26–28; Eitan Ben-Eliahu, "Air Force in the 21st Century," *Ma'arachot* 368 (March 2000), pp. 28–31 (Hebrew); Itai Brun, "Asymmetrical Warfare," *Ma'arachot* 371 (June 2000), pp. 44–47; Isaac Ben-Israel, "Technology and Decision: Some Thoughts for the IDF after Kosovo," *Ma'arachot* 371 (June 2000), pp. 34–43 (Hebrew).

75. Lanir, "From Operational Art," p. 11. On the paradigmatic shift in the IDF, also see: Reuven Gal, "The Israeli Defense Forces (IDF): A Conservative or an Adaptive Organization?" in Daniel Maman, Eyal Ben-Ari, and Zeev Rosenhek, *Military, State and Society in Israel* (London: Transaction, 2001), pp. 367–68; Shmuel Gordon, *The Bow of Paris: Technology, Doctrine and Israeli Security* (Tel Aviv: Sifriat Ha Poalim, 1997), pp. 306–10.

76. Naveh, "Cult of Offensive Preemption," pp. 180–81, and "New Elements of Maneuver," pp. 20–29; Lanir, "From Operational Art," pp. 13–14; Weizman, "Walking through Walls," p. 8, n. 13; Glick, "Halutz's Stalinist Moment"; Doron Minart, "From Armor to Air: Mechanization," *Ma'arachot* 370 (April 2000), pp. 2–15 (Hebrew).

77. Zeev Schiff, "Operational Revolution in the IDF," *Ha'aretz*, 7 September 2004; Amir Rappoport, "Reformulating the Victory," *Ma'ariv* (May 2004); Barzilai, "IDF Prepared for Changed Warfare," "Getting the Aftermath Right," and "Should We Be Up in Arms over Egypt's Buildup?" *Ha'aretz*, 18 January 2005; Ofer Shelah and Yoav Limor, *Captives in Lebanon* (Tel Aviv: Yediot Ahronot Books, 2007), p. 197 (Hebrew); *The Commission of Inquiry into the Events of Military Engagement in Lebanon 2006* (Tel

Aviv: Winograd Commission, hereafter *Winograd Report*), p. 48; Alon Ben David, "All Quiet on the Eastern Front, so Israel Will Revise IDF Organization and Doctrine," *Jane's International Defense Review* (March 2004), p. 51, and "The IDF Adapts Doctrine and Structure in Response to Shifting Regional Priorities," *Jane's International Defense Review* (March 2005); *IDF Strategy* (TOHAD, July 2001). For Mofaz's vision, see: Shaul Mofaz, "The IDF in the 2000s," *Ma'arachot* 363 (March 1999), pp. 2–9 (Hebrew).

78. Craig Dalton, *Systemic Operational Design: Epistemological Bumpf or the Way Ahead for Operational Design?* (Fort Leavenworth, KS: SAMS, 2006); Matt Matthews, *We Were Caught Unprepared: The 2006 Hezbollah–Israeli War* (Fort Leavenworth, KS: CSI Press, 2008), pp. 24–25.

79. Schiff, "Operational Revolution"; Weizman, "Lethal Theory," p. 60; Shelah and Limor, *Captives in Lebanon*, pp. 198–200; Ya'ari and Assa, *Diffused Warfare*, pp. 30, 59; Barzilai, "Should We Be Up in Arms?"; Jonathan Marcus, "Israel Sharpens Its Military Strategy," BBC News Special Report, 29 April 1998; Ya'acov Zigdon, "Fire Loops: Directions in Force Buildup and Fire Capabilities," *Ma'arachot* 386 (November 2002), pp. 10–15 (Hebrew).

80. Zeev Bonen, "The Influence of PGMs on Ground Battles," *Ma'arachot* 375–76 (February 2001), pp. 60–65 (Hebrew); Douglas Davis, "IDF Reform: Proposals Leaked to London Newsletter," *Jerusalem Post*, (22 April 1999); Udi Shani, "The Grand Vision of Communications in the IDF," *Ma'arachot* 400 (May 2005), pp. 14–19 (Hebrew).

81. Barzilai, "Getting the Aftermath Right"; Eyido Hecht, "Decision Mechanisms: How to Win the War?" *Ma'arachot* 385 (September 2002), pp. 4–19 (Hebrew); Gershon Hacohen, "Test of a Result as a Test of Tendency," *Ma'arachot* (May 2002), pp. 12–16 (Hebrew).

82. *Winograd Report*, p. 49; Barzilai, "Getting the Aftermath Right"; Gad Sneh, "Striating Epistemology with Operations: Effects-Based Synchronization," presented at Transformation in Warfare: Effects Based Operations conference, Tel Aviv, 2002; *Effect-Based Operations: Concepts and Doctrines in the World Militaries—the USA* (Tel Aviv: TOHAD, 2003) (Hebrew); Lt. C. Sharon, "The Information Era and the Chemistry of Warfare," *Ma'arachot* 403-4 (December 2005), pp. 46–49 (Hebrew); Eran Ortal, "The GS Officer in the Era of Jointness," *Ma'arachot* (June 2005), pp. 28–31 (Hebrew); Alex Fishman, "The IDF in His Hands," *Yediot Ahronot*, 26 January 2007, p. 15; Barbara Opall-Rome, "Rehabilitating Israel's Land Forces," *Defense News*, 7 May 2007.

83. Col. A., "Use of the Air Force in the Second Gulf War," *Ma'arachot* 390 (July 2003), pp. 16–29 (Hebrew); Isaac Ben-Israel, "The Manifestation of the RMA in the Second Gulf War," in *In the Wake of the War in Iraq*, ed. Shai Fieldman and Moshe Grundman (Tel Aviv: Jaffe Center for Strategic Studies, 2004), pp. 69–91 (Hebrew); Amnon Barzilai, "Iraq War Convinced the IDF: It Is Time for the Technological Revolution," *Ha'aretz*, 6 May 2003 (Hebrew).

84. Seth G. Jones, "Fighting Networked Terrorist Groups: Lessons from Israel," *Studies in Conflict and Terrorism* 30 (2007), pp. 281–302; Ya'ari and Assa, *Diffused Warfare*; Gal Hirsch, "On Dinosaurs and Hornets: A Critical View on Operational Moulds in Asymmetric Conflicts," *RUSI Journal* (August 2003); Shelah and Limor, *Captives in Lebanon*, p. 198; Weizman, "Lethal Theory," pp. 61–62, 63–64, and "Walking through Walls"; Felix Frish, "Thanks to the Plasma," *Ma'ariv*, 23 February 2007, pp. 19–21.

85. Barzilai, "Tank of the Future"; Chris Demchak, "Technology's Knowledge Burden, the RMA and the IDF: Organizing the Hypertext Organization for Future Wars of Disruption," in *Israel's National Security towards the 21st Century*, ed. Uri Bar-Joseph (London: Frank Cass, 2001), pp. 77–146; Lorie Jewell, "Unified Quest 04 Focuses on Joint Capabilities," *Army News Service* (10 May 2004).

86. Lanir, "From Operational Art," p. 11.

87. *Jointness: The Concept of Future Force Operation* (Tel Aviv: IDF/TOHAD, 2002); Barzilai, "Getting the Aftermath Right" and "Tank of the Future"; *Concept of Future Joint Forces Operations* (TOHAD, October 2002) (Hebrew); *Joint Operations in Urban Environments* (TOHAD, March 2005) (Hebrew); Barbara Opall-Rome, "Remote Controlled Military Guards Israeli Borders," *Defense News*, 4 April 2005; Ortal, "GS Officer"; Zvi Lanir, "Why the Jointness Concept Is Needed," *Ma'arachot* 401 (September 2005), pp. 20–27 (Hebrew); Ariel Granit, "Joint Experiments in the USA," *Ma'arachot* 378–79 (September 2001), pp. 66–71 (Hebrew).

88. Barzilai, "Getting the Aftermath Right"; Lanir, "From Operational Art," pp. 10–11.

89. Meir Finkel, "Flexible Force Structure: A Flexibility-Oriented Force Design and Development Process in Israel," *Israel Affairs* 12, no. 4 (October 2006), pp. 789–800; Udi Adam, "Transformation from Sectarian Logistics to Combined Arms Techno-Logistics," *Ma'arachot* 400 (May 2005), pp. 9–14 (Hebrew).

90. *Quadrennial Defense Review 2001* (Washington, DC: DoD, Office of the Secretary of Defense, 30 September 2001).

91. Sergio Catignani, *Israeli Counter-Insurgency and the Intifadas* (London: Routledge, 2008), pp. 10270.

92. Moshe Ya'alon, "Preparation for Low-Intensity Conflict," *Ma'arachot* (December 2001), pp. 380–81 (Hebrew).

93. Hirsch, "On Dinosaurs" and "The Development of the Campaign in the Central Military District, 2000–2003," in Haggai Golan and Shaul Shay, Low Intensity Conflict (Tel Aviv: Ma'arachot, 2004), pp. 239–51; Golan and Shay, *Low-Intensity Conflict*; Tamir Eshel, "LIC-2005 Focuses," *Military Technology* 3 (2005), pp. 98–101.

94. Sergio Catignani, "Israel Defense Forces Organizational Changes in an Era of Budgetary Cutbacks," *RUSI Journal* (October 2004), and "The Strategic Impasse in Low-Intensity Conflicts: The Gap between Israeli Counter-Insurgency Strategy and Tactics during the Al-Aqsa Intifada," *Journal of Strategic Studies* (2005); Barbara

Opall-Rome, "A Wider View: Israeli Firm Networks," *Defense News*, 31 January 2005, and "Virtual Occupation," *C4ISR Journal* 32 (March 2006); Robin Hughes and Alon Ben David, "Double Jeopardy: Israel Is Caught in the Crossfire of Two Extreme Threats to Its Security," *Jane's Defense Weekly* (17 November 2004), and "Interview with Major General Eliezer Shkedy," *Jane's Defense Weekly* (5 January 2005); Alon Ben David, "Inner Conflict: Israel's Low-Intensity Conflict Doctrine," *Jane's Defense Weekly* (1 September 2004); Maj. Chen, "Interpreting Real-Time Intelligence From UAVs," *Ma'arachot* 394 (October 2004), pp. 41–47 (Hebrew); Lt. C. Eyal and Maj. Doron, "From Vision to Reality in Battle-Supporting Intelligence," *Ma'arachot* 373 (November 2002), pp. 33–37 (Hebrew).

95. Shelah and Limor, *Captives in Lebanon*, p. 200; Robin Hughes, "Interview with Lieutenant General Moshe Ya'alon," *Jane's Defense Weekly* (17 November 2004).

96. Alon Ben David, "Israel Equips First Division with DAP Capability," *Jane's Defense Weekly* (23 March 2005), and "Israel Closes Intelligence Gap," *Jane's Defense Weekly* (23 March 2005); Opall-Rome, "C4ISR Dominates Israeli Investment Focus" and "Israel Seeks to Extend Precise Ground Strike," *Defense News*, 12 September 2005, "Israel Builds Precision Arsenal for Ground War," *Defense News*, 13 April 2005, "Israeli QDR Emphasizes Multi-role, Linked Force," *Defense News*, 13 February 2006, and "Surgical Strike to the Highest Bidder," *Defense News*, 15 May 2006.

97. Avi Kober, "The Israeli Defense Forces in the Second Lebanon War: Why the Poor Performance?" *Journal of Strategic Studies* 31, no. 1 (2008), pp. 30–32; Stuart Cohen, *Israel and Its Army* (London: Routledge, 2008), pp. 104–6.

98. *Winograd Report*, pp. 407–8; Ofra Graicer, "Lessons for Doctrine and Force Structure," *The Second Lebanon War: Lessons for Modern Militaries*, RUSI Conference Proceedings, 20 June 2008; Amos Harel and Avi Issacharoff, *Spider Webs* (Tel Aviv: Yediot Ahronot, 2008) (Hebrew), pp. 111–21; Amira Lam, "The Shelf Blunder," *7 Days* (9 March 2007); Alon Ben David, "Debriefing Teams Brand IDF Doctrine Completely Wrong," *Jane's Defense Weekly* (3 January 2007); Ziv Maor, "Don't Blame the Plasmas, Blame the Training on Decentralized Warfare," *Omedia* (29 January 2007).

99. Shelah and Limor, *Captives in Lebanon*, p. 197.

100. Ibid., pp. 198–99; Matthews, *We Were Caught Unprepared*, p. 64.

101. Cohen, *Israel and Its Army*, p. 105.

102. Giora Segal, "Art of Tactical Warfare: Commanders in the Battlefield Chaos," *Ma'arachot* 398 (January 2005), pp. 8–21 (Hebrew). For the critique of incomprehensible orders by tactical commanders during the intifada, see: Roman Gofman, "Low-Intensity Conflict on the Low Tactical Level," *Ma'arachot* 390 (July 2003), pp. 66–67 (Hebrew); Lt. C. Raz, "The Dangerous Language of the Low-Intensity Conflict," *Ma'arachot* 380–81 (December 2001), pp. 54–55 (Hebrew); and during the second Lebanon war: Shelah and Limor, *Captives in Lebanon*, pp. 197–98; Rappoport, "Warning:

Swarms with Teeth," pp. 7–11; Yaakov Katz, "IDF Think Tank Chief: Better Tactical Language Needed," *Jerusalem Post*, 20 December 2006.

103. Yehuda Wagman, "Low Intensity: The Failure," pp. 251–99 in Golan and Shay, *Low-Intensity Conflict*; Ya'acov Amidror, "Operational Options or Rationale?" *Ma'arachot* 388 (April 2003), pp. 4–8 (Hebrew).

104. Dan Fauytkin, "Post-Modernism and Military Thought," *Ma'arachot* 406 (April 2006), pp. 16–21 (Hebrew); Avraham Zohar, "The Emperor Has No Clothes," *Ma'arachot* 386 (November 2002), pp. 64–66 (Hebrew).

105. Shai Shabtai "The Concept of Dynamic Molecule," *Ma'arachot* 397 (October 2004), pp. 78–79; Shmuel Gordon, "There Is No Place for Blind Admiration," *Ma'arachot* 392 (December 2003), pp. 63–67 (Hebrew).

106. Ya'acov Amidror, "The Strike as Paradigm of Cognitive Effects," *Ma'arachot* 403–4 (December 2005), pp. 54–57 (Hebrew); Yehuda Wagman, "What Was the Operational Goal?" *Ma'arachot* 403–4 (December 2005), pp. 58–69 (Hebrew).

107. *Winograd Report*, pp. 267–74, 407–48; Harel and Issacharoff, *Spider Webs*, pp. 112–14, and *The Seventh War* (Tel Aviv: Yediot Ahronot, 2004), pp. 58–59; Matthews, *We Were Caught Unprepared*, pp. 26–27; Shelah and Limor, *Captives in Lebanon*, pp. 198–99; Ron Tira, *The Limitations of Standoff Firepower Based Operations*, INSS Memo no. 89, 2007; Stephen Biddle, *The 2006 Lebanon Campaign and the Future of Warfare* (Carlisle, PA: US Army War College, Strategic Studies Institute, 2008).

108. Raymond Cohen, *Culture and Conflict in Egyptian–Israeli Relations: A Dialogue of the Deaf* (Bloomington: Indiana UP, 1990), pp. 28–29; Motti Regev, "To Have a Culture of Our Own: On Israeliness and Its Variations," *Ethnic and Racial Studies* 23, no. 2 (2000), pp. 223–47.

109. Raymond Cohen, "Culture Gets in the Way," *Middle East Quarterly* 1, no. 3 (September 1994).

110. Baruch Kimmerling, *The Invention and Decline of Israeliness: State, Society and the Military* (Berkeley: U of California P, 2001), pp. 56–130; Lucy Shahar and David Kurz, *Border Crossings: American Interactions with Israelis* (Yarmouth, ME: Intercultural Press, 1995), pp. 72–73; Miriam Erez and Christopher Earley, "Comparative Analysis of Goal-Setting Strategies across Cultures," *Journal of Applied Psychology* 72 (1987), pp. 658–65, and *Culture, Self-Identity, and Work* (New York: Oxford UP, 1993); Elihu Katz and Tamar Liebes, "Decoding Dallas: Notes from a Cross-Cultural Study," in *Television: The Critical View*, ed. Horace Newcomb (New York: Oxford UP, 1987), pp. 419–32.

111. Shahar and Kurz, *Border Crossings*, p. 73. See, for example: Yair Galily and Ken Sheard, "Cultural Imperialism and Sport: The Americanization of Israeli Basketball," *Culture, Sport, Society* 5, no. 2 (Summer 2002), pp. 55–78.

112. Geert Hofstede, *Culture's Consequences: International Differences in Work-Related Values* (Beverly Hills, CA: Sage, 2001), p. 457; Rebecca Merkin, "Power Distance and Facework Strategies," *Journal of Intercultural Communication Research* 35,

no. 3 (2006), pp. 139–60; Hadas Wiseman and Amia Lieblich, "Individuation in a Collective Community," *Adolescent Psychiatry* 18 (1992), pp. 156–79.

113. Hofstede, *Culture's Consequences*, pp. 90, 121; Judith Martin and Thomas Nakayma, *Intercultural Communication in Contexts* (New York: McGraw-Hill, 2006), pp. 239–40; Shahar and Kurz, *Border Crossings*, pp. 97, 102–5.

114. Cohen, Eisenstadt, and Bacevich, *Knives, Tanks, and Missiles*, pp. 65–66.

115. Tamar Kartiel, *Talking Straight: Dugri Speech in Israeli Sabra Culture* (Cambridge: Cambridge UP, 1986); Aharon Klieman, "Israeli Negotiating Culture," in *How Israelis and Palestinians Negotiate*, ed. Tamara Cofman Wittes (Washington, DC: United States Institute for Peace Press, 2005), p. 94; Oz Almog, *The Sabra: A Profile* (Tel Aviv: Am Oved, 1997) (Hebrew), pp. 230–31; Shahar and Kurz, *Border Crossings*, p. 79.

116. Almog, *Sabra*, pp. 229–30, 372; Klieman, "Israeli Negotiating Culture," pp. 94–95; Shahar and Kurz, *Border Crossings*, pp. 115–16, 127, 144–45; Kartiel, *Talking Straight*.

117. Danolad Ellis and Ifat Maoz, "Dialogue and Cultural Communication Codes between Israeli Jews and Palestinians," in Larry Samovar, Richard Porter, and Edwin McDaniel, *Intercultural Communication* (New York: Thomson Wadsworth, 2006), pp. 231–38; Alean Al-Krenawi, John Graham, and Mueen Fakher-Aldin, "Telephone Counseling: A Comparison of Arab and Jewish Israeli Usage," *International Social Work* 46, no. 10 (2003), pp. 495–509.

118. Cohen, *Negotiating across Cultures*, p. 124, *Culture and Conflict in Egyptian–Israeli Relations*, and "Meaning, Interpretation and International Negotiation," *Global Society* 14, no. 3 (2000), pp. 317–35; Shahar and Kurz, *Border Crossings*, pp. 70, 77–78, 137.

119. Cohen, *Culture and Conflict in Egyptian–Israeli Relations*; Klieman, "Israeli Negotiating Culture," ch. 4, and esp. p. 94.

120. Shahar and Kurz, *Border Crossings*, pp. 96–99, 146.

121. Amos Handel, "Cognitive Styles among Adolescents in Israel," *International Journal of Psychology* 8, no. 4 (1973), pp. 255–67; Zachary Dershowitz, *Influences of Cultural Patterns on the Thinking of Children: A Study of the Effect of Jewish Subculture on the Field Dependence–Independence Dimension of Cognition* (PhD diss., New York University, 1966); B. Zakik, "Field Dependence–Independence among Oriental and Western Schoolchildren," *Megamot: Behavior Sciences Quarterly* 16 (1968), pp. 51–58 (Hebrew); Shlomo Sharan and Leonard Weller, "Articulation of the Body Concept among First-Grade Israeli Children," *Child Development* 42 (1971), pp. 1553–59, and "Classification Patterns in Unprivileged Children in Israel," *Child Development* 42 (1971), pp. 581–94; Sara Druyan and Iris Levin, "Differential Contribution of Field-Dependent and Field-Independent Cognitive Styles on Socio-cognitive Transaction in Promoting Scientific Reasoning," *International Journal of Behavioral Development* 19, no. 4 (1996), pp. 831–50; Naomi Katz, "Individual Learning Styles: Israeli Norms and Cross-Cultural Equivalence of Kolb's Learning Style Inventory," *Journal of Cross-Cultural Psychology* 19, no. 3 (1988), pp. 361–79; Baruch Margalit and Paul

Mauger, "Cross-Cultural Demonstration of Orthogonality of Assertiveness and Aggressiveness: Comparison between Israel and the United States," *Journal of Personality and Social Psychology* 46, no. 6 (1984), pp. 1414–21; Ilana Preale, Yehuda Amir, and Shlomo Sharan, "Perceptual Articulation and Task Effectiveness in Several Israel Subcultures," *Journal of Personality and Social Psychology* 15, no. 3 (1970), pp. 190–95. Also, for field dependence–independence and holistic versus analytical cognition, see: Tamar Globerson and Tamar Zelniker, *Cognitive Style and Cognitive Development* (Norwood, NJ: Ablex, 1989).

122. "He (Moshe) took the Book of the Covenant and read it in earshot of the people, and they said, 'Everything that Hashem has said, we will do and we will obey'" (*Naase ve nishma*) (Shemot 24: 7), http://www.askmoses.com/article.html?h=412&o=52046.

123. Charles D. Freilich, "National Security Decision Making in Israel: Processes, Pathologies, and Strengths," *Middle East Journal* 60, no. 4 (Autumn 2006), p. 661.

124. Klieman,"Israeli Negotiating Culture," p. 95.

125. Cohen, Eisenstadt, and Bacevich, *Knives, Tanks, and Missiles*, p. 68.

126. Elena Grigorenko, Robert Sternberg, and Sidney Strauss, "Practical Intelligence and Elementary-School Teacher Effectiveness in the United States and Israel: Measuring the Predictive Power of Tacit Knowledge," *Thinking Skills and Creativity* 1 (2006), pp. 14–33.

127. Cohen, Eisenstadt, and Bacevich, *Knives, Tanks, and Missiles*, p. 125.

128. Michael Handel, "The Evolution of Israeli Strategy: The Psychology of Insecurity and the Quest for Absolute Security," in Williamson Murray, Alvin Bernstein, and MacGregor Knox, *The Making of Strategy: Rulers, States and War* (Cambridge: UK: Cambridge UP, 1996), p. 534.

129. For the first time in history, Israel formulated its national security concept in written form in spring 2006, when the Committee Charged with Formulation of National Security Conception under Dan Meridor submitted its report. Zeev Schiff, "Security Concept Should Be Updated," *Ha'aretz*, 2 August 2004; "Report: Additional Countries in the Middle East Will Procure Nukes," *Ha'aretz*, 24 April 2006; Yuval Azulai, "National Security Concept Report," *Ha'aretz*, 31 May 2006; Alon Ben David, "Israel Outlines Defense Doctrine," *Jane's Defense Weekly* (3 May 2006); Amir Rappoport, "Territories Do Not Have Security Importance Any More," *Ma'ariv*, 2 September 2005), and "New Security Concept," *Ma'ariv*, 7 May 2006); Ofir Shelach, "The Shelf Will Be Very Happy," *Yediot Achronot*, 28 April 2006.

130. Steven J. Rosen and Martyn Indyk, "The Temptation to Preempt in the Fifth Arab–Israeli War," *Orbis* 20, no. 3 (Summer 1976); David Rodman, "Patterns of War Initiation in the Arab–Israeli Conflict: A Note on the Military Dimension," *Israel Affairs* 6, no. 3/4 (Spring/Summer 2000), pp. 115–26.

131. Bar-Joseph, ed., *Israel's National Security*, pp. 2–5, "The Paradox of Israeli Power," *Survival* 46, no. 4 (Winter 2004–5), pp. 137–56, and "The Conceptualization of Deterrence in Israeli Thinking," *Survival* 7, no. 3 (1998), pp. 145–81; David Ben-Gurion, *Goal and Specific: Thoughts about Israeli Security* (Tel Aviv: Ma'arachot, 1971) (Hebrew), pp. 145, 175–89, 193–95, 200–213, 242, 251; Handel, "The Evolution of Israeli Strategy"; Mark Heller, *Continuity and Change in Israeli Security Policy* (Oxford: Oxford UP, 2000); Zeev Maoz, *Defending the Holy Land: A Critical Analysis of Israeli Security and Foreign Policy* (Ann Arbor: U of Michigan P, 2006); Efraim Inbar, "Israel's National Security in the Twenty-First Century," *Israel Affairs* 12, no. 4 (October 2006); David Rodman, *Defense and Diplomacy in Israel's National Security Experience* (Portland, OR: Sussex Academic Press, 2005).

132. Handel, "The Evolution of Israeli Strategy," p. 570; Clive Jones, "A Reach Greater than the Grasp: Israeli Intelligence and the Conflict in South Lebanon 1990–2000," *Intelligence and National Security* 16, no. 3 (Autumn 2001), pp. 4–6.

133. Naveh, "Cult of Offensive Preemption," pp. 168–87; Ariel Levite, *Israeli Military Doctrine: Offense and Defense* (Tel Aviv: Jaffe Center for Strategic Studies, 1988) (Hebrew); Handel, "The Evolution of Israeli Strategy," p. 537; David Rodman, "Combined Arms Warfare in the Israeli Defense Forces: An Historical Overview," *Defense Studies* 2, no. 1 (Spring 2002), pp. 109–26, and "Patterns of War"; Eduard Luttwak and Dan Horowitz, *The Israeli Army* (New York: Harper and Row, 1975), pp. 63, 108; Emanuel Wald, *The Wald Report: The Decline of Israeli National Security since 1967* (Oxford: Westview Press, 1992), p. 97; Stuart Cohen and Efraim Inbar, "A Taxonomy of Israel's Use of Force," *Comparative Strategy* 10, no. 2 (April 1991), pp. 121–38.

134. Israel Tal, *The Few against the Many* (Tel Aviv: Ma'arachot, 1998), pp. 68–85; Cohen, Eisenstadt, and Bacevich, *Knives, Tanks, and Missiles*, pp. 23–24; Rodman, "Israel's National Security Doctrine: An Appraisal," p. 127; Yehuda Wagman, "When the IDF Knew to Solve Problems," *Makor Rishon*, 4 June 2006.

135. *Wald Report*, p. 97; Baruch Kimmerling, "The Social Construction of Israel's National Security," in Stuart Cohen, *Democratic Societies and Their Armed Forces: Israel in Comparative Context* (London: Frank Cass Publishers, 2000), pp. 228–33; Isaac Ben-Israel, *The Missiles' War* (Tel Aviv: Tel Aviv University, 2007), p. 33 (Hebrew); Naveh, "Cult of Offensive Preemption," pp. 169–71.

136. Yaniv, *Politics and Strategy*, p. 299; Naveh, "Cult of Offensive Preemption," pp. 169–71.

137. Ben-Gurion, *Goal and Specific*, p. 13. Cohen, Eisenstadt, and Bacevich, *Knives, Tanks, and Missiles*, p. 59; Ruth Bondy, Ohad Zmora, and Raphael Bashan, *Mission Survival* (New York: Sabra Books, 1968), pp. 120–21.

138. Zvi Ofer and Avi Kober, eds., *Quality and Quantity in Military Buildup* (Tel Aviv: Ma'arachot, 1985) (Hebrew).

139. Cohen, Eisenstadt, and Bacevich, *Knives, Tanks, and Missiles*, pp. 54–55, 67; Freilich, "National Security Decision Making," p. 663.

140. Ben-Israel, "Where Did Clausewitz Go Wrong?" pp. 16–26; Elyashiv Shimshi, *By Force of Stratagem* (Tel Aviv: Ma'arachot, 1995) (Hebrew); Meir Pail, "Stratagem, Motivation and Technology," in Ofer and Kober, *Quality and Quantity*, pp. 361–73; Gad Navon, "Conceptual Creativity and Motivation as Sources of Decision," in Ofer and Kober, *Quality and Quantity*, pp. 373–77; Cohen, Eisenstadt, and Bacevich, *Knives, Tanks, and Missiles*, pp. 55, 57. For examples of creativity in IDF military history, see: Shmuel Gordon, "Technological Warrior and the War in Lebanon," *Ma'arachot* 350 (January 1997), pp. 20–29 (Hebrew).

141. Cohen, Eisenstadt, and Bacevich, *Knives, Tanks, and Missiles*, p. 51. Brian Bond, *Liddell Hart: A Study of His Military Thought* (New Brunswick, NJ: Rutgers UP, 1977), pp. 268–73; Michael Handel, *Israel's Political Military Doctrine* (Cambridge, MA: Harvard UP, 1973), pp. 67–68; Luttwak and Horowitz, *Israeli Army*, pp. 62–64; Basil Liddell Hart, *The Memoirs of Captain Liddell Hart* (London: Cassel, 1965), vol. 2, p. 183. On the question of Liddell Hart's influence on Israeli military thought, see: Tuvia Ben-Moshe, "Liddell Hart and the Israel Defense Forces: A Reappraisal," *Journal of Contemporary History* 16 (1981), pp. 369–91.

142. Zeev Bonen, "The Smart Use of Technology," in Ofer and Kober, *Quality and Quantity*, pp. 409–13; Uri Dromi, "Technology and Military Systems," in Ofer and Kober, *Quality and Quantity*, pp. 433–47.

143. Moshe Dayan, *Milestones* (Jerusalem: Edanim, 1976), pp. 112–13; Luttwak and Horowitz, *Israeli Army*, pp. 108–9; Gilli Vardi, "Pounding Their Feet: Israeli Military Culture as Reflected in Early IDF Combat History," *Journal of Strategic Studies* 31, no. 2 (April 2008), p. 303; Cohen, Eisenstadt, and Bacevich, *Knives, Tanks, and Missiles*, p. 55; Yigal Alon, *Kilim Sheluvim* (Tel Aviv: Hakibbutz HaMeuhad, 1980) (Hebrew), p. 102.

144. Handel, "The Evolution of Israeli Strategy," p. 563.

145. Abraham Ben-Zvi, *The United States and Israel: The Limits of the Special Relationship* (New York: Columbia UP, 1993), *John F. Kennedy and the Politics of Arms Sales to Israel* (London: Frank Cass, 2002), and *In the Shadow of the Hawk: Lyndon B. Johnson and the Politics of Arms Sales to Israel* (London: Frank Cass, 2004). Also see Cohen, Eisenstadt, and Bacevich, *Knives, Tanks, and Missiles*, p. 59. For a somewhat different interpretation, see David Rodman, *Arms Transfers to Israel: The Strategic Logic behind American Military Assistance* (Portland, OR: Sussex Academic Press, 2007).

146. Handel, "The Evolution of Israeli Strategy," p. 549; Monia Meridor, *Rafael: On the Roads of Research and Development of Israeli Security* (Tel Aviv: Ministry of Defense Publishing House, 1981) (Hebrew); Stewart Reiser, *The Israeli Arms Industry: Foreign Policy, Arms Transfers and Military Doctrine of the Small State* (New York:

Holmes and Meier, 1989); Aharon Klieman and Reuven Pedatzur, *Rearming Israel: Defense Procurement through the 1990s* (Tel Aviv: Jaffe Center for Strategic Studies, 1992); Cohen, Eisenstadt, and Bacevich, *Knives, Tanks, and Missiles*, p. 44.

147. David Rodman, "Israel's National Security Doctrine: An Introductory Overview," *MERIA* 5, no. 3 (September 2001); Cohen, Eisenstadt, and Bacevich, *Knives, Tanks, and Missiles*, pp. 51, 53–54. Also see: Efraim Katzir, "Did Science Contribute to Israeli National Security?" *Ma'arachot* 363 (March 1999), pp. 44–53 (Hebrew).

148. Bonen, *Rafael*. See especially introduction by Shimon Peres. Handel, "The Evolution of Israeli Strategy," pp. 546, 550, 568–69; Freilich, "National Security Decision Making," p. 638.

149. Reuven Gal, *A Portrait of the Israeli Soldier* (Westport, CT: Greenwood, 1986), pp. 174–76; Chen Itzhaki, "Technology Is Not Everything," *Ma'arachot* (September 1993), pp. 2–7 (Hebrew); Ben-Israel, "Back to the Future," pp. 2–5, and "Technological Lessons," *Ma'arachot* (October 1993), pp. 8–13 (Hebrew); Barbara Tuchman, *Practicing History: Selected Essays* (New York: Ballantine Books, 1981), pp. 180–81.

150. Martin van Creveld, *Military Lessons of the Yom Kippur War: Historical Perspectives* (Beverly Hills, CA: Sage, 1975), pp. 2–3.

151. Handel, "The Evolution of Israeli Strategy," pp. 546, 559.

152. Cohen, Eisenstadt, and Bacevich, *Knives, Tanks, and Missiles*, p. 59.

153. Approaching problems with no military solution, Israel frequently relied on hard power and saw military operations "as choices of the first, rather than last, resort." Handel, "The Evolution of Israeli Strategy," pp. 555, 562–63. Isaacson, Layne, and Arquilla, *Predicting Military Innovation*, p. 29; Kimmerling, "Social Construction of Israel's National Security," pp. 215–54.

154. Uri Bar-Joseph, "Towards a Paradigm Shift in Israel's National Security Conception," *Israel Affairs* 6, no.2 (Spring 2000), p. 106.

155. *Wald Report*, p. 190.

156. Cohen, Eisenstadt, and Bacevich, *Knives, Tanks, and Missiles*, pp. 61, 68–69, 76.

157. Rodman, "Israel's National Security Doctrine," p. 126; Cohen, Eisenstadt, and Bacevich, *Knives, Tanks, and Missiles*, p. 61.

158. Klieman, "Israeli Negotiating Culture," p. 88; Wagman, "Service in the IDF: Is It a National Mission?" *Ma'arachot* (January 1999)," p. 8.

159. Cohen, Eisenstadt, and Bacevich, *Knives, Tanks, and Missiles*, pp. 81–82; Naveh, "Cult of Offensive Preemption," p. 177.

160. Handel, "The Evolution of Israeli Strategy," p. 546.

161. Daniel Bar-Tal, *Shared Beliefs in a Society: Social Psychological Analysis* (London: Sage, 2000), pp. 107, 111, 143–44; Klieman, "Israeli Negotiating Culture," pp. 85–86; *Wald Report*, p. 215; Avner Yaniv, "A Question of Survival: The Military and Politics under Siege," in Yaniv, *National Security and Democracy in Israel* (New York: Boulder, 1993); E. M. Bruner and P. Gorfain, "Dialogic Narration and the Paradoxes of

Masada," in E. M. Bruner, *Text, Play and Story* (Washington, DC: American Ethnological Society, 1983), pp. 56–79; Barry Schwartz, Yael Zerubavel, and Bernice M. Barnett, "The Recovery of Masada: A Study in Collective Memory," *Sociological Quarterly* 27, no. 2, pp. 147–64.

162. Gabriel Ben-Dor, "Regional Culture and the NACD in the Middle East," in Keith R. Krause, *Culture and Security* (London: Frank Cass, 1998), pp. 198–204, and "Responding to the Threat," pp. 113–38.

163. Freilich, "National Security Decision Making," p. 637; Bar-Tal, *Shared Beliefs in a Society*, pp. 98–99; Daniel Bar-Tal and Dikla Antebi, "Siege Mentality in Israel," *International Journal of Intercultural Relations* 16, no. 3 (1992), pp. 251–75, and "Beliefs about Negative Intentions of the World: A Study of the Israeli Siege Mentality," *Political Psychology* 13, no. 4 (1992), pp. 633–45; Ehud Luz, "Through the Jewish Historical Prism: Overcoming a Tradition of Insecurity," in Bar-Tal, Jacobson, and Klieman, *Security Concerns*, pp. 55–73; Yossi Beilin, *Israel: A Concise Political History* (New York: St. Martin's Press, 1992).

164. Handel, "The Evolution of Israeli Strategy," p. 534; Bar-Tal, *Shared Beliefs in a Society*, pp. 94–95.

165. Klieman, "Israeli Negotiating Culture," p. 90; Handel, "The Evolution of Israeli Strategy," pp. 542–43; Shahar and Kurz, *Border Crossings*, pp. 109, 147–48.

166. Bar-Tal, *Shared Beliefs in a Society*, pp. 93, 108–12; Nurith Gertz, "The Security Narrative in Israeli Literature and Cinema," in Bar-Tal, Jacobson, and Klieman, *Security Concerns*, pp. 193–215; Benyamin Neuberger and Ilan Ben-Ami, *Democracy and National Security in Israel* (Tel Aviv: Open University, 1997) (Hebrew).

167. Bar-Tal, *Shared Beliefs in a Society*, p. 96; Baruch Kimmerling and Moshe Lissak, *Military and Security* (Jerusalem: Hebrew University, 1996) (Hebrew); Yaniv, "A Question of Survival: The Military and Politics under Siege"; Yael Zerubavel, *Recovered Roots: Collective Memory and the Making of Israeli National Tradition* (Chicago: U of Chicago P, 1995).

168. Gregory Giles, *Continuity and Change in Israel's Strategic Culture* (McLean, VA: Defense Threat Reduction Agency, SAIC, 2006), p. 13; Efraim Inbar, "The 'No Choice War' Debate in Israel," *Journal of Strategic Studies* 23 (March 1989), pp. 22–37; Cohen, Eisenstadt, and Bacevich, *Knives, Tanks, and Missiles*, p. 55; Shahar and Kurz, *Border Crossings*, pp. 25–26; Freilich, "National Security Decision Making," p. 652.

169. Yaniv, *National Security and Democracy*, ch. 4; Raymond Cohen, "Israeli Intelligence before the 1956 Sinai Campaign," *Intelligence and National Security* (January 1988), pp. 100–140; Bar-Tal, *Shared Beliefs in a Society*, p. 95.

170. Bar-Joseph, "Towards a Paradigm Shift," pp. 104–7; *Wald Report*, pp. 193–247; Yehezkel Dror, "Israel's Quest for Ultimate Security: Strategies and Perceptions," in Bar-Tal, Jacobson, and Klieman, *Security Concerns*, pp. 433–47; Klieman, "Israeli Negotiating Culture," p. 90; Handel, "The Evolution of Israeli Strategy," pp. 542–43.

171. Handel, "The Evolution of Israeli Strategy," pp. 539–40.

172. Reuven Pedatzur, "Ben-Gurion's Enduring Legacy," in Bar-Tal, Jacobson, and Klieman, *Security Concerns*, pp. 139–67; Edna Lomsky-Feder, *The Military and Militarism in Israeli Society* (Albany: SUNY Press, 1999).

173. Aharon Klieman, "Israel: Succumbing to Foreign Ministry Declinism," in Brian Hocking, *Foreign Ministries: Change and Adaptation* (London: Macmillan, 1999), ch. 5; Klieman, "Israeli Negotiating Culture," p. 81.

174. Although true in certain respects, a siege mentality frequently distorted perceptions of the Israeli strategic culture. Cohen, Eisenstadt, and Bacevich, *Knives, Tanks, and Missiles*, p. 60; Nadav Safran, *Israel: The Embattled Ally* (Cambridge, MA: Harvard UP, 1978), pp. 48–51; Monia Meridor, *The Army on the Way to the Country* (Tel Aviv: Ma'arachot, 1988), pp. 200–20 (Hebrew); Gil Merom, "Israel's National Security and the Myth of Exceptionalism," *Political Science Quarterly* 114, no. 3 (1999), p. 431.

175. Klieman, "Israeli Negotiating Culture," p. 90.

176. Freilich, "National Security Decision Making," p. 643; Handel, "The Evolution of Israeli Strategy."

177. Stuart Cohen, "An Exchange on Israel's Security Doctrine," *MERIA* 5, no. 4 (December 2001), p. 128.

178. Handel, "The Evolution of Israeli Strategy," p. 570; Freilich, "National Security Decision Making," p. 646; *Wald Report*, p. 215; Vardi, "Pounding Their Feet," p. 312.

179. Freilich, "National Security Decision Making," pp. 647, 653–54.

180. Cohen, Eisenstadt, and Bacevich, *Knives, Tanks, and Missiles*, pp. 71–73; Dan Horowitz, "Flexible Responsiveness and Military Strategy: The Case of the Israeli Army," *Policy Science* (April 1970), pp. 191–205; Gal, "Israeli Defense Forces," pp. 367–68.

181. Freilich, "National Security Decision Making," pp. 646–49; Handel, "The Evolution of Israeli Strategy," pp. 559–62; *Wald Report*, p. 216; Giles, *Continuity and Change*, p. 17.

182. Eliezer Goldberg, "Particularistic Considerations and the Absence of Strategic Assessment in the Israeli Public Administration: The Role of the State Comptroller," *Israel Affairs* 11, no. 2 (Spring 2005), pp. 434–44.

183. Shahar and Kurz, *Border Crossings*, pp. 23, 68, 84–85.

184. Brent McKnight and Nick Bontis, "E-improvisation: Collaborative Groupware Technology Expands the Reach and Effectiveness of Organizational Improvisation," *Knowledge and Process Management* 9, no. 4 (2002), pp. 219–27; Ira Sharkansky and Yair Zalmanovitch, "Improvisation in Public Administration and Policy Making in Israel," *Public Administration Review* (July–August 2000), pp. 321–29.

185. Freilich, "National Security Decision Making," pp. 644, 653, 657, 660; Shahar and Kurz, *Border Crossings*, pp. 69–71, 85, 105–9, 120–21.

186. Klieman, "Israeli Negotiating Culture," p. 94; Luttwak and Horowitz, *Israeli Army*, pp. 87, 16–163, 174; Gal, *Portrait of the Israeli Soldier*, p. 130; Isaacson, Layne, and Arquilla, *Predicting Military Innovation*, pp. 25–26; Winer, "Who Needs the Doctrine?" p. 43.

187. *Wald Report*, p. 190; Isaacson, Layne, and Arquilla, *Predicting Military Innovation*, pp. 30–31; Luttwak and Horowitz, *Israeli Army*, p. 54; Cohen, Eisenstadt, and Bacevich, *Knives, Tanks, and Missiles*, p. 61.

188. Freilich, "National Security Decision Making," p. 662; Uzi Benziman, "It Is All Shoddy," *Ha'aretz*, 29 November 2006; Shahar and Kurz, *Border Crossings*, pp. 72, 133–34.

189. Freilich, "National Security Decision Making," pp. 644, 651–55; Yehuda Ben Meir, *National Security Decision Making: The Israeli Case*, JCSS Study no. 8 (Tel Aviv: Jaffe Center for Strategic Studies, 1986), pp. 67–69, 84–91.

190. Freilich, "National Security Decision Making," p. 644; Gal, *Portrait of the Israeli Soldier*, p. 10; Cohen, Eisenstadt, and Bacevich, *Knives, Tanks, and Missiles*, p. 49.

191. Handel, "The Evolution of Israeli Strategy," p. 571; *Wald Report*, p. 83; Vardi, "Pounding Their Feet," pp. 310–22; Zohar, "The Emperor Has No Clothes," p. 64; Gal, *Portrait of the Israeli Soldier*, p. 14; *Agranat Report: National Commission of Inquiry Appointed to Investigate the Circumstances Leading to the Outbreak of the Yom Kippur War* (Tel Aviv: Agranat Commission, 1974), summary chapter, pp. 1464–65; Ari Shavit, "There Is a Limit," *Ha'aretz*, 14 September 2007.

192. Gal, *Portrait of the Israeli Soldier*, p. 115.

193. Ibid., p. 130; Winer, "Who Needs the Doctrine?" p. 45; Shahar and Kurz, *Border Crossings*, p. 119.

194. *Wald Report*, pp. 116–19; Gal, *Portrait of the Israeli Soldier*, p. 171; Yigal Alon, *The Making of the Israeli Army* (New York: Bantam Books, 1970), pp. 11–12; Benziman, "It Is All Shoddy."

195. Martin van Creveld, *The Sword and the Olive* (New York: Public Affairs Books, 2002), p. 161.

196. Vardi, "Pounding Their Feet," pp. 295–309.

197. In many ways this attitude served as a replacement for the traditional expression *be'ezrat Ha-Shem* (with God's help), which expressed a passive wishful thinking. Almog, *Sabra*, pp. 181–83; Shahar and Kurz, *Border Crossings*, p. 70.

198. Amihad Shahar, "The Problem Is Not in the IDF," *Ma'arachot* (December 2001), pp. 88–89; Almog, *Sabra*, pp. 181–83.

199. Wagman, "Service in the IDF"; *Wald Report*; Freilich, "National Security Decision Making"; Winer, "Who Needs the Doctrine?" p. 43; Dany Reshef, "What Is Wrong with the Beseder Culture?" *Ma'arachot* 361 (November 1998), pp. 90–92 (Hebrew).

200. *Wald Report*, p. 166.

201. Cohen, Eisenstadt, and Bacevich, *Knives, Tanks, and Missiles*, p. 79.

202. Gal, "Israeli Defense Forces," p. 363; Shahar and Kurz, *Border Crossings*, p. 70.

203. Shahar and Kurz, *Border Crossings*, pp. 12, 115–16, 125–26; Almog, *Sabra*, pp. 124–29, 209–11, 216–35.

204. Almog, *Sabra*, pp. 124–29, 209–11, 216–35.

205. Cohen, Eisenstadt, and Bacevich, *Knives, Tanks, and Missiles*, p. 50; Shahar and Kurz, *Border Crossings*, p. 71; Raymond Cohen, "Israel's Starry-Eyed Foreign Policy," *Middle Eastern Quarterly* 1, no. 2 (June 1994).

206. Leo Rosten, *The New Joys of Yiddish* (New York: Crown, 2001).

207. Freilich, "National Security Decision Making," pp. 653, 659; Yigal Alon, *Kilim Sheluvim*; Almog, *Sabra*, pp. 333, 423.

208. Shahar and Kurz, *Border Crossings*, pp. xvi, 121; Almog, *Sabra*, pp. 220–26.

209. Almog, *Sabra*, pp. 124–29, 209–11, 216–35; Gertz, "Security Narrative," pp. 193–215; Yehuda Wallach, *Establishment of the People's Army during the War* (Tel Aviv: Ma'arachot, 1997), p. 118 (Hebrew). "Fool's hat," which symbolized, among other things, neglect to intellectualism, became a national cultural symbol. Almog, *Sabra*, pp. 241–42; Editorial, "Pardon My Turkish," *Jewish Daily Forward*, 23 October 2003; Mira Liond, "The Journey after the Tembel," *Kol-Bo* 33 (27 April 1990), p. 35.

210. Ya'akov Hasdai, "Bitsuist and Ideologue: The Priest and the Prophet of the IDF," *Ma'arachot* 280 (May 1981), pp. 41–49 (Hebrew), and "Social and Political Revolution Is Needed," *Neemanei Tora ve Avoda* 10 (October 2006) (Hebrew); Winer, "Who Needs the Doctrine?"; Wallach, *Establishment of the People's Army*, pp. 165–69; Eviatar Ben-Zedef, "The IDF Is Not a Professional Army," *News First Class* (October 2006), p. 16 (Hebrew).

211. Yahiam Witz, *Palmach: Sheaves and Sword* (Tel Aviv: Ministry of Defense Publishing House, 2000) (Hebrew); Nahum Bogner, *Military Thought in the Hagana* (Tel Aviv: Misrad ha Bitahon, 1998) (Hebrew); Cohen, Eisenstadt, and Bacevich, *Knives, Tanks, and Missiles*, p. 53; Ya'acov Bengo, "The Strategic Missile and Infantry Platoon," *Ma'arachot* 400 (May 2005), pp. 96–97 (Hebrew).

212. Yoav Gelber, *The Core of the Regular Army* (Tel Aviv: Yad ben Tzvi, 1984) (Hebrew); Zehava Ostfeld, *The Army Is Born* (Tel Aviv: Ministry of Defense, 1994) (Hebrew); Meir Pail, *From Defense to Defense Forces* (Tel Aviv: Zmora Beitan, 1979), esp. pp. 38–40 (Hebrew); Wallach, *Establishment of the People's Army*, pp. 112, 117–18; Winer, "Who Needs the Doctrine?" p. 42; Zohar, "The Emperor Has No Clothes," pp. 64–65; Bar-Joseph, "Towards a Paradigm Shift," p. 99; Cohen, Eisenstadt, and Bacevich, *Knives, Tanks, and Missiles*, pp. 53, 79.

213. Winer, "Who Needs the Doctrine?" p. 43.

214. Kober, "Intellectual and Modern Focus," pp. 151–56; Isaacson, Layne, and Arquilla, *Predicting Military Innovation*, pp. 24–25.

215. Vardi, "Pounding Their Feet," pp. 295–324.

216. Ibid., pp. 295–308.

217. Kober, "Intellectual and Modern Focus," p. 152; Tal, *Few against the Many*.

218. *Wald Report*, p. 13; Bar-Joseph, *Israel's National Security*, p. 1; Cohen, Eisenstadt, and Bacevich, *Knives, Tanks, and Missiles*, p. 75.

219. *Wald Report*, pp. 12, 119–20.

220. Winer, "Who Needs the Doctrine?" p. 43.

221. Cohen, Eisenstadt, and Bacevich, *Knives, Tanks, and Missiles*, pp. 50, 65–66, 76; Meir Pail, *The Commander: Proper Ways for Military Leadership* (Tel Aviv: Hakibbutz HaMeuhad, 2003), pp. 11–29 (Hebrew).

222. Ya'acov Amidror, "The Sources of Lack of Knowledge in Military Affairs," *Ma'arachot* 378–79 (2001)," p. 23; Gal, *Portrait of the Israeli Soldier*, pp. 190–209.

223. Shahar, "Problem Is Not in the IDF," p. 89; Ya'acov Tzur, "Did the IDF: Learn and Implement the Lessons from Its Campaigns?" *Ma'arachot* 378–79 (2001)," p. 17; Wagman, "Service in the IDF," p. 13.

224. Cohen, Eisenstadt, and Bacevich, *Knives, Tanks, and Missiles*, pp. 49, 74–76.

225. *Wald Report*, pp. 13–14, 208–9.

226. Kober, "Intellectual and Modern Focus," pp. 151–56.

227. Wagman, "Service in the IDF," p. 12; Kober, "Intellectual and Modern Focus."

228. Emanuel Wald, *The Owl of Minerva* (Tel Aviv: Yediot Ahronot, 1994) (Hebrew), pp. 36–37; Cohen, Eisenstadt, and Bacevich, *Knives, Tanks, and Missiles*, p. 75; Kober, "Intellectual and Modern Focus."

229. Wagman, "Service in the IDF," p. 11; *Agranat Report*, p. 283.

230. Kober, "Intellectual and Modern Focus," p. 143; Winer, "Who Needs the Doctrine?" p. 45; Boaz Zalmanovich, "IDF Commander's Reading List," *Ma'arachot* 418 (April 2008), pp. 58–61.

231. Kober, "Intellectual and Modern Focus," pp. 144–45, 153.

232. Isaacson, Layne, and Arquilla, *Predicting Military Innovation*, pp. 24–25; Kober, "Intellectual and Modern Focus," pp. 151–56.

233. Yehuda Wagman, "Service in the IDF p. 8 (Hebrew); Avriel Bar-Levav, *War and Peace in the Jewish Culture* (Jerusalem: Merkaz Zalman Shazar, 2006), pp. 11–67, 223–47, 313–49 (Hebrew); Avraham Shapira, *Conversation of the Fighters* (Tel Aviv: Ma'arechet ha-Ihud, 1967), pp. 8–9 (Hebrew); Gertz, "Security Narrative," pp. 95–96, 103–6. Also see Stuart Cohen, "The Scroll or the Sword? Tension between Judaism and Military Service in Israel," in Cohen, *Democratic Societies and Their Armed Forces: Israel in Comparative Context* (London: Frank Cass, 2000), pp. 254–74.

234. Cohen, *Israel and Its Army*, pp. 19–34.

235. Anita Shapira, *The Sword of the Dove* (Tel Aviv: Am Oved, 1992), pp. 9–11, 489–90 (Hebrew); Almog, *Sabra*, p. 216; Yigal Alon, *The Display of Sand* (Tel Aviv: Hakibbutz HaMeuhad, 1959), pp. 272–73 (Hebrew).

236. Alon, *The Display of Sand*, p. 314; Wagman, "Service in the IDF," pp. 8–10.

237. Michael Handel, *Masters of War: Classical Strategic Thought* (New York: Routledge, 2000), p. 82.

238. Cohen, Eisenstadt, and Bacevich, *Knives, Tanks, and Missiles*, p. 74.

239. Ibid.

240. *Agranat Report*, summary chapter, pp. 1464–65.

241. Ben-Gurion, *Goal and Specific*, p. 175.

242. Wagman, "Service in the IDF," p. 12; Cohen, Eisenstadt, and Bacevich, *Knives, Tanks, and Missiles*, p. 138.

243. StuartCohen, *Israel and Its Army: Change and Continuity* (New York: Routledge, 2007), p. 101.

244. Alon, *Display of Sand*, p. 303; Cohen, Eisenstadt, and Bacevich, *Knives, Tanks, and Missiles*, p. 75; Gal, *Portrait of the Israeli Soldier*, p. 170.

245. Almog, *Sabra*, pp. 226–27; Alon, *Display of Sand*, p. 303.

246. Ben-Israel, *Missiles' War*, p. 38; Amos Harel, "State Comptroller: Many IDF Senior Officers Poorly Trained," *Ha'aretz*, 4 December 2006; Reuven Gal, "Testing the Current Model of the Israeli Officer," *Ma'arachot* 346 (February 1996), pp. 18–25; Richard A. Gabriel and Reuven Gal, "The IDF Officer: Linchpin in Unit Cohesion," *Army* 34, no. 1, pp. 42–50.

247. Ya'acov Amidror, "The Sources of Lack of Knowledge," pp. 20–24, "What Is Professional Military Knowledge?" *Ma'arachot* 372 (2000), pp. 27–31, and "The Essence of the Military Profession," *Ma'arachot* 369 (2000); Gabriel and Gal, "IDF Officer," pp. 42–50; Winer, "Who Needs the Doctrine?" p. 43.

248. Wagman, "Service in the IDF," p. 11.

249. *Wald Report*, pp. 12–13, 129.

250. Amidror, "Sources of Lack of Knowledge"; also see: Dany Davidi, "The Influence of the Tactical Command College on the Officers Corps in the IDF," *Ma'arachot* 396 (September 2004), pp. 32–35 (Hebrew); Eyal Ben-Reuven, "The Challenge of Officers' Education in the IDF," *Ma'arachot* 396 (September 2004), pp. 6–10 (Hebrew).

251. Emanuel Wald, *The Owl of Minerva* (Tel Aviv: Yediot Ahronot, 1994), pp. 31–48 (Hebrew); Kober, "Intellectual and Modern Focus," pp. 144–45, 153; Wagman, "Service in the IDF," p. 11; Ben-Moshe, "Liddell Hart," pp. 374–75; Wallach, *Establishment of the People's Army*, pp. 130–36.

252. Cohen, *Israel and Its Army*, p. 102.

253. Cohen, Eisenstadt, and Bacevich, *Knives, Tanks, and Missiles*, p. 74; Stuart Cohen and Ilan Suleiman, "The IDF: From People's to Professional Army," *Armed Forces and Society* (1995), pp. 237–51.

254. Cohen, *Israel and Its Army*, pp. 103–4; Lieutenant Colonel Muli, "The Command and Staff College: Military Academy," *Ma'arachot* 347 (April 1996); Uzi Lev-Tzur, "The Program of Officers Training in the Military College," *Ma'arachot* 361 (May 1989); Gal, *Portrait of the Israeli Soldier*, p. 125; Moshe Shamir, "On the Changes in the Combined Arms Command and Staff College," *Ma'arachot* 396 (September 2004), pp. 20–25 (Hebrew).

255. Gal, *Portrait of the Israeli Soldier*, pp. 170–71; Cohen, Eisenstadt, and Bacevich, *Knives, Tanks, and Missiles*, p. 74.

256. Amidror, "Sources of Lack of Knowledge"; Amos Harel, "State Comptroller: Many IDF Senior Officers Poorly Trained," *Ha'aretz*, 4 December 2006; Cohen, Eisenstadt, and Bacevich, *Knives, Tanks, and Missiles*, pp. 67, 74–75.

257. Vardi, "Pounding Their Feet," p. 309.

258. Ya'acov Tzur, "Did the IDF," pp. 12–19.

259. *Wald Report*, pp. 117–18.

260. Amidror, "Sources of Lack of Knowledge," p. 21.

261. Raymond Cohen, "Living and Teaching across Cultures," *International Studies Perspectives* 2 (2001), pp. 151–60.

262. Avigdor Kahalani, *A Warrior's Way* (New York: Shapolsky, 1994), pp. 263, 266–67; Gal, *Portrait of the Israeli Soldier*, p. 171.

263. Rodman, "Israel's National Security Doctrine: An Introductory Overview," see the *self-reliance* section; Cohen, Eisenstadt, and Bacevich, *Knives, Tanks, and Missiles*, p. 52.

264. Isaac Ben-Israel, "The Logic of Learning Military Lessons," *Ma'arachot* 305 (1986) (Hebrew).

265. Tzur, "Did the IDF," p. 18.

266. Yehuda Ben Meir, *National Security Decision Making: The Israeli Case*, JCSS Study no. 8 (Tel Aviv: Jaffe Center for Strategic Studies, 1986), pp. 84–91; Amir Bar-Or, "The Link between the Government and the IDF during Israel's First 50 Years: The Shifting Role of the Defense Minister," in Daniel Maman, Eyal Ben-Ari, and Zeev Rosenhek, *Military, State and Society in Israel* (London: Transaction Publishers, 2001), pp. 321–43; Freilich, "National Security Decision Making," pp. 651, 657–59, 642.

267. Michael, "Israel Defense Forces," pp. 421–46.

268. Cohen, Eisenstadt, and Bacevich, *Knives, Tanks, and Missiles*, p. 78; Yehuda Ben-Meir, *Civil Military Relations in Israel* (New York: Columbia UP, 1995).

269. Bar-Tal, *Shared Beliefs in a Society*, p. 147; Wagman, "Service in the IDF," p. 12; Kimmerling, "Social Construction of Israel's National Security," p. 224; Micha Popper, "The Israeli Defense Forces as a Socializing Agent," in Bar-Tal, Jacobson, and Klieman, *Security Concerns*, pp. 167–81; Nisim Solomon, "Education in the IDF," *Ma'arachot* (July 1982); Dan Horowitz and Moshe Lissak, *Trouble in Utopia: The Overburdened Polity of Israel* (Albany: SUNY Press, 1989), pp. 195–230; Michael Curtis and

Mordechai Chertoff, *Israel: Social Structure and Change* (New York: Transaction Books, 1973), pp. 419–32.

270. Israel Tal, *The Few against the Many* (Tel Aviv: Ma'arachot, 1998), p. 95; *Wald Report*, pp. 165–66.

271. *Wald Report*, pp. 164–65; Wallach, *Establishment of the People's Army*, pp. 146–52.

272. Ortal, "GS Officer"; *Wald Report*, pp. 182–83.

273. Tal, *Few against the Many*, pp. 11, 89, 95–97; *Wald Report*, pp. 181–82, 187–89.

274. *Wald Report*, pp. 170–71, 186–87, 189–90; Editorial, "The Army Has No Commander," *Ha'aretz*, 15 December 2006.

Conclusion

1. Margaret E. Keck and Kathryn Sikkink, *Activists beyond Borders* (Ithaca, NY: Cornell UP, 1996); Ethan A. Nadelmann, "Global Prohibition Regimes: The Evolution of Norms in International Society," *International Organization* 44, no. 4 (1990), pp. 480–525; Ann Marie Clark, *Diplomacy of Conscience: Amnesty International and Changing Human Rights Norms* (Princeton, NJ: Princeton UP, 2002); Emanuel Adler, "Seizing the Middle Ground: Constructivism in World Politics," *European Journal of International Relations* 3, no. 3 (1997); Ron Eyerman and Andrew Jamison, *Social Movements: A Cognitive Approach* (University Park, PA: Pennsylvania State UP, 1991); Peter Haas, "Introduction: Epistemic Communities and International Policy Coordination," *International Organization* 46, no. 1 (special issue, Winter 1992).

2. Theo Farrell, *The Norms of War: Cultural Beliefs and Modern Conflict* (London: Lynne Reinner Publishers, 2005), p. 13, and esp. ch. 2.

3. Stephen Peter Rosen, *Winning the Next War: Innovation and the Modern Military* (Ithaca, NY: Cornell UP, 1994).

4. Thomas Mahnken, *Technology and the American Way of War* (New York: Columbia UP, 2008), pp. 24, 27, 28–55.

5. P. M. Derevianko, *Revoliutsiia v voennom dele: vchem ee sushchnost'?* (Moscow: Ministerstvo Oborony SSSR, 1967); Andrei Kokoshin, *O revoliutsii v voennom dele v proshlom I nastoiashem* (Moscow: Lenannd, 2006), pp. 24–26; V. A. Zolotarev, *Istoriia voennoi strategii Rossii* (Moscow: Kuchkovo Pole, 2000), pp. 377–448.

6. Avner Cohen, "Nuclear Arms in Crisis under Secrecy: Israel and the Lessons of the 1967 and 1973 Wars," in *Planning the Unthinkable: How New Powers Will Use Nuclear, Biological, and Chemical Weapons*, ed. Peter Lavoy, Scott Sagan, and James Wirtz, (Ithaca, NY: Cornell UP, 2000), pp. 104–25; *Hatabu haaharon* (Kinneret: Zmora Beitan, 2005), "Crossing the Threshold: The Untold Nuclear Dimension of the 1967 Arab–Israeli War and Its Contemporary Lessons," *Arms Control Today* 37, no. 5 (June 2007), pp. 12–16, and "Going for the Nuclear Option," *Ha'aretz*, 23 May 2007; Yuval Ne'eman, "The USA–Israel Connection in the Yom Kippur War," in *Nuclear Weapons and the 1973*

Middle East War, ed. Michael Wheeler and Kemper Gay, Center foi National Security Negotiations, Occasional Paper, Nuclear Lessons and Legacies Project, Monograph no. 3 (1996); Shmuel Tzabag, *Yehudim, Israel, ve astrategiia gari'init* (Ariel: Ehuda ve Shomron, 2007).

7. Peter Boyer, "Downfall," *New Yorker*, 20 November 2006, p. 4.

8. H. R. McMaster, "On War: Lessons to be Learned," *Survival* 50, no. 1 (February 2008), p. 19, and "Learning from Contemporary Conflicts to Prepare for Future War," *Orbis* 52, no. 4 (Fall 2008), pp. 564–84.

9. Barry Watts and Williamson Murray, "Military Innovation in Peace Time," in Williamson Murray and Allan Millet, *Military Innovation in the Interwar Period* (Cambridge: Cambridge UP, 1996), pp. 404–6.

10. Robert M. Gates, "Remarks to the Heritage Foundation," Colorado Springs, Colorado, 13 May 2008, US Dept. of Defense, Office of the Asst. Sec. of Defense (Public Affairs), http://www.defenselink.mil/speeches/speech.aspx?speechid=1240.

11. Michael Horowitz and Dan Shalmon, "The Future of War and American Military Strategy," *Orbis* 53, no. 2 (Spring 2009), pp. 301–2.

12. Jacqueline Newmyer, "Filling a Gap in Our Understanding of Chinese Strategy," Testimony before the US–China Economic and Security Review Commission Hearing on Chinese Military Modernization and US Export Controls, 16 March 2006, and "Oil, Arms and Influence: The Indirect Strategy behind Chinese Military Modernization," *Orbis* 53, no. 2 (Spring 2009), pp. 205–19.

13. Colin Gray, *Out of the Wilderness: Prime Time for Strategic Culture* (Fort Belvoir, VA: Defense Threat Reduction Agency, 2006), p. 17.

14. Uri Bar-Joseph and Rose McDermott, "The Intelligence Analysis Crisis," in *The Oxford Handbook of National Security Intelligence*, ed. Loch Johnson (New York: Oxford UP, 2010); James Witz, "The American Approach to Intelligence Studies," in *Handbook of Intelligence Studies*, ed. Loch Johnson (London: Routledge, 2009); Austin Yamada, "Counterintelligence and US Strategic Culture," in *Vaults, Mirrors and Masks*, ed. Jennifer Sims and Burton Gerber (Washington, DC: Georgetown UP, 2009).

15. For example, see: Assaf Moghadam, *The Globalization of Martyrdom: Al Qaeda, Salafi Jihad, and the Diffusion of Suicide Attacks* (Baltimore: Johns Hopkins UP, 2008); Brynjar Lia and Thomas Hegghammer, "Jihadi Strategic Studies: The Alleged Al Qaida Policy Study Preceding the Madrid Bombings," *Studies in Conflict and Terrorism* 27, no. 5 (September–October 2004), pp. 355–75; Jeni Mitchell, "The Contradictory Effects of Ideology on Jihadist War Fighting: The Bosnia Precedent," *Studies in Conflict and Terrorism* 31, no. 9 (September 2008), pp. 808–28.

16. Patrick Porter, *Military Orientalism: Eastern War through Western Eyes* (New York: Columbia UP, 2009).

17. Stephen Peter Rosen, "After Proliferation: What to Do If More States Go Nuclear," *Foreign Affairs* 85, no. 5 (September 2006); Andrei Kokoshin, "Nuclear Conflict

in the Twenty-first Century," discussion paper 2007-03, Belfer Center for Science and International Affairs, Harvard University, April 2007.

18. Williamson Murray, "Does Military Culture Matter?" *Orbis*, 43 (Winter 1999), p. 41.

19. Max Boot, *War Made New: Technology, Warfare and the Course of History, 1500 to Today* (New York: Gotham Books, 2006), p. 9.

20. Emily Goldman and Leslie Eliason, *Diffusion of Military Technology and Ideas* (Stanford, CA: Stanford UP, 2003), pp. 6–10.

INDEX

Active Defense, 155*n*11
adaptation, 21
adaptive learning, 134–35
Advanced Operational Course (KOMEM), 100, 101; curriculum, 101
Advanced Operational Group, 99
Afghanistan, 39
Air-Land Battle (ALB), 60, 61, 100; formal release, 26; neutralizing, 26; origination of, 25–26; sophistication, 176*n*25
Amiel, Saadia, 94
analytical attention, 19
annihilation by fire, 78–79, 91, 133
anthropological psychology, 16
anti-intellectual: Israeli Defense Forces, 118, 119–22, 128; Israeli national cultural symbol, 216*n*209; United States, 80
anti-Semitism, 115
Arab-Israeli conflict, 115
arms race, linear, 61
Assault-Breaker, 35, 60, 178*n*41; threat, 63
a-strategic thinking, 79–81
asymmetrical threats, 13
authoritarian power, 48–50, 171*n*162
automated information processing, 28–29

Bar-Joseph, Uri, 114, 139
Bathurst, Robert, 77
Bekaa Valley, 94–95
Ben-Dor, Gabriel, 115
Ben-Israel, Isaac, 124
Ben-Zvi, Abraham, 113
Berlin, Isaiah, 41, 46

Blitzkrieg, 112
Boot, Max, 145*n*3; on innovation, 1
Bracken, Paul, 71
Brodie, Bernard, 15

carrier warfare, 146*n*17
casualties: Soviet tolerance, 44; United States avoidance, 76, 86
China, 139; cognitive style, 153*n*15
Civil War, US, 79
Clausewitz, Carl von, 82
cognitive continuum, 18–20
cognitive style, 9, 16–23, 23; China, 153*n*15; Israeli, 110–11, 134; national, 15; Revolution in Military Affairs, 20–23; Soviet Union, 39–42, 133–34, 153*n*15; United States, 75–78, 134
Cohen, Eliot, x, 79, 86, 111
Cold War: learning through historical analogies, 141; military balance assessments, 66; Soviet triple transformation during, 29; theory of victory, 137–38; United States military effectiveness in post, 175*n*7
collective memoirs, 7
collectivistic society, 17; Israel transformed away from, 97; Soviet Union, 39, 164*n*94
combined-systems revolutions, 75
command, control, communications, computers, and information (C4I), 2
Command Staff College, 123
Commission on Integrated Long-Term Strategy, 66–69

Soviet Union: abstract ideas, 52–53; Active
Defense, 155*n*11; approach to studying
war, 41–42; asymmetric response, 46;
attitude towards environment, 173*n*187;
authoritarian tradition, 48–50, 171*n*162;
avant-gardism in art/poetry, 169*n*146;
avoidance of reality, 172*n*181; belief in
survival in nuclear war, 158*n*30;
casualties tolerance, 44; ceremonies v.
quest for truth, 164*n*94; cognitive style,
39–42, 133–34, 153*n*15; collapse, 179*n*49,
194*n*232; collectivistic society, 39;
collision of East/West, 164*n*94;
commonness with people, 39; computer
production, 52; creativity/improvisation,
171*n*161; cultural characteristics, 39–42;
cultural contradictions, 164*n*94;
declaring MTR, 72–73; deep operations
battle, 33; deep stand-off striking ability,
31; deep targeting, 34; defense
procurement, 46; defensive
counteroffensive, 30; defensive
sufficiency, 38; despotism, 164*n*94;
direction for operational surprise,
171*n*160; doctrine, 44–46; echelonment
doctrine, 25–26; endurance, 42;
envisioning enemy as system, 42;
ethnocentrism, 51; extreme physical
conditions, 42; failure in
implementation of MTR, 37–39, 53;
"forecast and foresight," 48; forecasting,
contradictory philosophical approach
to, 171*n*160; high-context culture, 39–40;
holistic-dialectical thought, 40, 54–55;
imagined future, 51; implications of
MTR, 33–37; individualism, 163*n*90;
innovations in 1920s, 169*n*146;
institutionalized military science, 132;
integrated battlefield, 35; intellectual
atmosphere, 132–33; intellectuals state of
mind, 41; intellectual tradition, 165*n*101;
irrationalism, 51; Israeli Defense Forces
criticized in terminology of, 199*n*41;
Iug-80 exercise, 36; make-believe,
172*n*181; Marshall's understanding of
MTR, 72; matching NATO capabilities,
64; matching weapons to practical
needs, 46; mathematicians, 153*n*17;
mechanization, 44; messianic search for
God v. militant godlessness, 164*n*94;
military science, 46–47; military
science, futuristic innovations in,
169*n*146; military science, political
interference in, 170*n*150; military
theoreticians of 1920s, 47;
multidisciplinary science, 41; mysticism,
153*n*17; nonlinear warfare, 39; nuclear
strategy, 136; numerical supremacy, 64;
"occurrence," 40; ONA assessment of
MTR, 67; open source account of MTR,
181*n*77; Operational Research Theory
Institute borrowing from lexicon of,
100; patience, 42, 44; perestroika, 38;
personality consciousness v. impersonal
collectivism, 164*n*94; pessimistic
impotence, 173*n*187; polychronic, 40;
pure offensive attack, 30; reassessment
of nuclear-conventional balance, 29;
reconstructing operational logic of
Western doctrine, 32; reticence toward
techno-euphoria, 44–45, 56; romantic
culture of warfare, 43–44; science,
166*n*103, 166*n*105; scientifically
organized society, 46–47; "scientific
forecast," 48; scientific inquiry, 41;
scientific reconstruction, 46–47;
scientific-technical revolution, 28,
33; seeking universal principles, 41;
self-conscious conceptualization, 46;
serfdom, 43; social reorganization in
agriculture, 169*n*146; Soiuz-83 exercise,
36; spiritual power, 43, 51; strategic
culture, 39–53; strategic defense, 30;
strategic mentality, 40; strategic
offensive, 30; system of systems, 35;
technological inferiority, 39, 43–44, 54;
technology, 44–46; temperament
contradictions, 164*n*94; theater
maneuvers, 31, 67; theater-strategic
operation, 34; theoretical writings on
tactics, 174*n*200; time orientation, 50–53,
172*n*178; tolerance, 42; top-down
organization, 48–50; triple
transformation during Cold War, 29;
uncontrolled improvisation, 171*n*160;
United States intelligence misinterpreting
MTR of, 62–64; unpredictability of life,
42; violence v. humanity, 164*n*94; war
culture, 42–44; Warsaw Pact 1978
exercise, 29–30; Warsaw Pact 1982
exercise, 30; way of war, 56; weapons
procurement, 44–46; Western Active

190n184; poor performance in Vietnam, 60; positive historic narrative, 87–88; power restrictions, 78; precision strike, 180n64; preference for overwhelming blow, 79; productive capacity, 78, 80; "pursuit of happiness," 81–82; rediscovery of campaign, 176n15; reductionist problem-solving, 82; representing "new beginning," 81; resorting to force, 190n184; romantic engineering, 81–82; social institutions, 77; space warfare, 180n64; strategic culture, 75–88, 79; superiority of, proof of, 194n232; symbolism of machine in, 86; technocentric warfare bias, 85; techno-euphoria, 74, 91; technological focuses, 60; technological romanticism, 85–87, 91; technology-driven mentality, 68; time perception, 82–83, 172n178; valuing practice at expense of theory, 77, 80; weapons acquisition policy, 82–83. *See also* Joint Chiefs of Staff, US; Office of Net Assessment, US

US Field Manual 100-5, 25
"unity of will," 152n11
Ustinov, Dmitry, 37

Varennikov, V. I., 37
Vietnam: measuring through body counts/ track kills, 76; United States poor performance in, 60
Voennaia Mysl' (VM), 31–32, 85; tactics, 53; translations, 62–63

Voennyi vestnik (VV), 31; techno-tactical issues, 53

Wald Report, 96
Waltz, Kenneth, 11
warfare: carrier, 146n17; diffused, 106–7; high-precision, 28; industrial, 79, 80, 133; information, 3, 67–68; irregular, 138; logistic, 78–79; Operation Desert Storm example of sophisticated conventional, 97; romantic culture of, 43–44; Soviet Union nonlinear, 39; space, 180n64; United States technocentric bias, 85
Warner, Edward, 180n66
War of Attrition, 114
Warsaw Pact (WP), 27; 1978 exercise, 29–30; 1982 exercise, 30
Watts, Barry, 74
Whitehead, Stuart, 183n100
Wilmot, Chester, 79
Winthrop, John, 87
Wohlstetter, Albert, 65, 66; introducing discontinuities, 69; on resources in war, 187n149
Wolf, Charles, 66
World War II: mirror imaging, 194n235; Soviet tactics in, 53; United States conduct during, 79

Ya'alon, Moshe, 102–3

Zarubezhnoe voennoe obozrenie (ZVO), 31